Short- and Long-Term Changes in Climate

Volume II

Author
Felix G. Sulman, M.D., D.V.M.
Head, Bioclimatology Unit
Hebrew University-Hadassah Medical Center
Jerusalem, Israel

CRC Press, Inc.
Boca Raton, Florida

Library of Congress Cataloging in Publication Data

Sulman, Felix Gad, 1907-
  Short and long term changes in climate.
  Bibliography: v. 1, p.   ; v. 2, p.
  Includes indexes.
  1. Climatic changes. 2. Bioclimatology.
  3. Paleoclimatology, I. Title.
  QC981.8.C5S93   551.6   81-15472
  ISBN 0-8493-6420-5 (v. 1)   AACR2
  ISBN 0-8493-6421-3 (v. 2)

   Direct all inquiries to CRC Press, Inc., 2000 Corporate Blvd., N.W., Boca Raton, Florida, 33431.

© 1982 by CRC Press, Inc.

International Standard Book Number 0-8493-6420-5 (Volume I)
International Standard Book Number 0-8493-6421-3 (Volume II)

Library of Congress Card Number 81-15472
Printed in the United States

PREFACE

The contents of this book, *Short- and Long-Term Changes in Climate,* are the result of interdisciplinary studies carried out by a 15-man team which has worked together for 20 years. We would like in particular to mention the devoted collaboration of Drs. M. Assael (Psychiatry), A. Danon (Pharmacology), S. Dikstein (Pharmacology), N. Hirschmann (Biochemistry), Y. Kaplanski (Endocrinology), Y. Koch (Endocrinology), D. Levy (Electronics), A. Lewy (Statistics), L. Lunkan (Electronics), I. Nir (Pharmacology), Y. Pfeifer (Biology), B. Shalita (Biology), E. Superstine (Pharmacy), E. Tal (Pharmacology), J. Tannenbaum (Electronics), and C. P. Weller (Pharmacology).

The topic of biometeorology and paleoclimatology is so fascinating that new research students are attracted to it every year, and the circle is growing steadily. This is, in fact, part of a much more extensive research field which has culminated in the development of a Bioclimatology Unit in our department. It all started from the senior author's experience and interest in diseases of man and animals and the wish to alleviate their sufferings. Thus, pharmacology and therapeutics became our main field of work since 1934. However, the exigencies of changing climates in Israel demanded our daily attention.

Our work in this sphere was initiated by a generous grant from the Florina Lasker Fund for the Research of Man in the Holy Land, administered by Professor Kalman J. Mann, Director of the Hadassah Medical Organization who is still the chief promoter of our work. Our research was also assisted by Professor A. Kreiser, Associate Dean of our Pharmacy School. During the course of the research, the U.S. Department of Health, Education and Welfare, through its branch of Environmental Health Services, helped our studies by Agreement # 06-005-3, administered by Dr. A. Henschel, Cincinnati. At a later stage the U.S. Office of Aerospace Research helped us with a considerable grant. Then, during the period of recession we were supported by the Amcor-Amron Co. of Tel-Aviv through the kind offices of Mr. S. Goldman. Special thanks are due to Dr. Y. Pardo and his wife, Hanita, who took it upon themselves to help us with advice and resources. For the last several years our Bioclimatology Unit has enjoyed the magnanimous patronage of Mr. Herman and Mrs. Elsie Lane, New York, whose generous help has made the present work possible.

Special thanks are due to Miss Yocheved Sussmann, who has been editing and proofreading our papers for many years past, and has now devoted again her untiring effort to putting this book into proper shape.

Mrs. Sylvia Farhi deserves our admiration and gratitude for having carried out the difficult work of typing the manuscript with great zeal and devotion.

Finally, we wish to thank Mr. B. J. Starkoff, President of CRC Press, Inc. and his staff, especially Mrs. Lisa Levine Eggenberger, for all their invaluable advice and guidance.

As this monograph contains mainly some guiding references connected with the impact of climate changes, the reader who wishes to have a biometeorological survey is referred to the standard books in this field: S. W. Tromp, *Medical Biometeorology* (1963 and 1980)[1] and *Progress in Biometeorology* (1963 and 1973).[2] The controversial topic of paleoclimatology has been exhaustively described by H. H. Lamb in *Climate — Present, Past and Future* (1972 and 1977).[3] German readers will find valuable information in the survey *Biometeorologie* by V. Faust (1977),[4] in the monograph *Unsichtbare Umwelt* by Herbert L. Koenig and S. Lang (1977),[5] and in H. Flohn, *Klimaschwankungen* (1963, 1969).[6]

Recent climate events, such as droughts in Africa and India, and a decline in rainfall affecting many other countries, have emphasized that climate can change rapidly and such changes can become important to human affairs and migration of nations. This is the message conveyed to us by the paleoclimatology chapters in Volume II.

Weather is something one has to live with. In many countries it is a universal topic of conversation — and rightly so. John Heywood, 1497—1580, a contemporary of Henry VIII, known as a dramatist and an epigrammatist, wrote a charming comedy "Play of the Weather". He called it "A New and Very Mery Enterlude of All Manner Wethers". The farce describes a worried "Jupiter" who has been implored to abolish the wanton machinations of "Saturne" producing cold, "Phebus" sending heat, "Eolus" governing the winds, and "Phebe" launching the rain. Hearing the complaints of eight witnesses, Mery-Reports the Vyce, Gyntylman, Marchaunt, Ranger, Water Myller, Wynde Myller, Gentylwomen, and Lannder, he soon learns that everybody wishes another type of weather to suit himself. Jupiter feels that he cannot give in to the whimsical demands of each claimant and, after much litigation, decides that the case should be dismissed with costs. Thus, everything returned to the status quo and has remained so ever since.

Professor F. G. Sulman, M.D., D.V.M.
Bioclimatology Unit
Hebrew University-Hadassah Medical Center
Jerusalem, Israel, 1981

INTRODUCTION

Scope of Climate Research

In recognition of the manifold ramifications of climate research the International Society of Biometeorology was founded in 1956 with headquarters in Oegstgeest (Leiden), The Netherlands, under the secretariat of Dr. Solco W. Tromp. It now consists of over 500 scientists from varying disciplinary backgrounds from over 50 countries. Within the society there are nine permanent study groups devoted to the following topics:

1. Effects of Heat and Cold in Animals and Man.
2. Effects of Weather and Climate on Human Health and Disease.
3. Effects of Weather and Climate on Animal Disease.
4. Effects of Weather and Climate on Plants and Trees.
5. Architectural, Urban, and Engineering Biometeorology.
6. Biological Effects of Natural Electric, Magnetic, and Electromagnetic Fields.
7. Physical, Physiological, and Therapeutic Effects of Ionized Air and Electroaerosols.
8. Biological Rhythms with Special Reference to Environmental Influences.
9. Physicochemical and Biological Fluctuating Phenomena.

Out of this awe-inspiring list the topics listed under numbers 1, 2, 5, 6, 7, 8, and 9 will be dealt with in the present monograph.

REFERENCES

1. **Tromp, S. W.,** *Medical Biometeorology,* Elsevier, Amsterdam, 1963; *Biometeorology,* Heyden & Sons, London, 1980.
2. **Tromp, S. W. and Bouma, J. A.,** *Progress in Biometeorology,* Vol. 1—4, Swets & Zeitlinger, Amsterdam, 1963, 1970, 1973.
3. **Lamb, H. H.,** *Climate — Present, Past and Future,* Vol. 1 and 2, Methuen, London, 1972, 1977.
4. **Faust, V.,** *Biometeorologie,* Hippokrates Verlag, Stuttgart, 1977.
5. **Koenig, H. L. and Lang, S.,** *Unsichtbare Umwelt,* 2nd ed., Technische Universitat, Munich, 1977.
6. **Flohn, H.,** *Klimaschwankungen,* Westdeutscher Verlag, Koln-Opladen 1963; *Klimanderungen,* Bonner Meteorologische Abhandlungen, Universitat Bonn, 1969, 1973.

THE AUTHOR

Professor Felix Gad Sulman was born and educated in Berlin. He is presently Professor Emeritus of Applied Pharmacology at the Hebrew University of Jerusalem. He holds doctorates of Medicine and of Veterinary Medicine from the University of Berlin, Germany. His specialties include Pharmacology, Endocrinology, and Laboratory Sciences in which he has worked since 1930.

He is a member of many scientific societies including the Society for Endocrinology and the Royal Society of Medicine, U. K.; the Endocrine Society, Society for Experimental Biology and Medicine, and the New York Academy of Sciences, U. S.; the International Society of Biometeorology and the American Institute of Medical Climatology.

Dr. Sulman is the author or co-author of over ten monographs. In addition, he has published over 500 scientific papers in collaboration with his assistants, and is an Honorary Member of the Sociedad Argentina de Farmacologia y Terapéutica.

DEDICATION

The research and its presentation in this book was made possible by the untiring and devoted help and advice of my wife, Edith, without whom the work could not have been undertaken.

SHORT- AND LONG-TERM CHANGES IN CLIMATE

Author
Felix G. Sulman, M.D., D.V.M.

TABLE OF CONTENTS

Chapter 3
Long-Term Climate Changes 73

Chapter 1

MAN'S REACTION TO SHORT-TERM CLIMATE CHANGES

I. MEDICAL IMPACT OF WEATHER CHANGES

A. Stress Reactions

1. Personality Changes

There is no doubt that heat changes the "temper" of man. Our research has shown that this is due to adrenal stimulation followed by exhaustion.[1]   Evidence from our studies supports the general conclusion that it is a mistake to think of heat as having either a general or a specific effect on performance, or indeed of having any single effect. Heat appears capable of affecting many different kinds of performance, including those depending principally on perceptual activities, on thought processes of differing complexity, and on sensorimotor coordination of response mechanisms. The nature of the effects of heat on performance will depend on the intensity and duration of the heat and on the particular combination with air electricity, especially ions, sferics, and electrofields.[2]

2. Stress Reaction

In assessing the problems of an American or European citizen who goes to the tropics, it is a fact that his life differs in many ways — social, economic, and cultural — from that of his compatriots who stay at home. This is due to the impact of environmental stress. In a number of studies, these nonclimatic differences in living conditions have had effects on health and efficiency data which probably outweighed any effects of heat. These general aspects of tropical living also tend to vitiate the use of individuals as their own controls for assessing heat effects. Comparisons are sometimes made between data collected after differing lengths of stay in the tropics or before and after proceeding on "home leave". Tropical residence of long duration increases an individual's sensitivity not only to heat, but to all climatic changes.

One index of the general well-being and efficiency of personnel in industrial or military organizations is the rate of reporting minor injuries and illness. A relationship between temperature and skin disease, minor complaints, loss of sleep, and lowered performance has been observed for men living in air-cooled and in conventionally ventilated compartments during tropical cruises; the sickness returns from ships in tropical waters have been reported as double those in temperate areas. Interestingly enough the same holds for the population of Israel of which a considerable part has been brought up in the temperate zone.

3. Special Research

The exceptional climate of Israel offers the study of man in three completely different mesoclimates: the Mediterranean climate which is extremely hot and humid in summer, especially on the coastline; the subtropical hill climate which offers all the extremes of the moderate climate in winter with the addition of dry heat in spring, summer, and autumn; the tropical dry climate of the Jordan Valley which is extremely exigent.

Research carried out by us has shown that a prolonged stay in all three climates may induce "tropical lethargy" due to adrenal exhaustion. Up to five different parameters of adrenal functions may be impaired by the different exigencies of the prevailing climate. Thus we found an impairment of adrenal cortex hormones as well as adrenal medullary hormones as the cause of heat exhaustion: diminished androcorticosteroids

(17-KS), glucocorticosteroids (17-OH), mineralcorticosteroids (aldosterone), epineph-rine (adrenaline), or norepinephrine (noradrenaline) turned out to be the cause of the complaints raised against living in a hot climate. The well-founded sufferings of the resident of the tropics can today be diagnosed by laboratory tests and specifically treated by adrenal cortex hormones, anabolic hormones, or MAO inhibitors, as re-quired.[2,10a]

4. Stress Adaptation

Traits of our constitution are firmly based on the individual genes inherited by us from our parents, which form our genotype (type inherited through genes). Experience has shown, however, that the genotype is not fixed; it can be changed by the environ-mental stress into a new phenotype (type created by external circumstances). The stress changes of the phenotype may form our character, our temperament, and our psyche. Yet they can be overthrown by a stress reaction (Figure 1), as shown by Selye.[3] Stress reactions may be caused by pain, intoxication, emotion, infection, physical trauma, and last but not least, by the weather and its electrical impacts.

Selye has rightly pointed out that every stress would evoke three stages of an adap-tation syndrome:

1. Stage 1 — Alarm reaction (acute) involving hypothalamus, pituitary, and espe-cially adrenal medulla and adrenal cortex.
2. Stage 2 — Resistance (subchronic) involving the same organs, a reaction experi-enced by young people, especially males, and inciting increased action of the aforementioned glands to cope with the challenge of the stress.
3. Stage 3 — Exhaustion (chronic) also involving the abovementioned organs, a reaction experienced in particular by females as well as old people or old-timers living for extensive periods in hot climates. This includes people living in a Foehn area of Europe, a Santa Ana area of California, or Sharav area of the Middle East.

The explanation for the difference in sex reaction can be derived from the fact that the male sex hormones (testosterone and adrenosterone) are our stress hormones. Be-cause of their chemical structure they are called 17-Ketosteroids (17-KS) and in men they amount to an average of 18 mg/day in the urine, or which 12 mg derive from the adrenal gland and 6 mg from the testes. In women urinary 17-KS rate only 12 mg/day being derived from the adrenal gland alone. This makes the female adrenal gland much more vulnerable to exhaustion from coping with stress and explains why women are more sensitive to climatic stress than men.

5. The Vagotonic Stress Reaction

There is no doubt that vagotonic people exist and that they suffer from their pro-pensity to show a strong reaction of their autonomous vagus nerve when under stress. This can occur even on rising in the morning, when they have to face the exigencies of daily life. The vagotonic patient shows all the reactions evoked by the nervous trans-mitter of the vagal system: acetylcholine.[4] He has an immediate reaction of all his salivary glands, a dripping nose, a tendency to vomit or to cough, loose bowels, a permanent movement of all the smooth muscles, especially those of the intestines, the bladder, and the gallbladder. His three vital parameters of life, i.e., blood pressure, pulse rate, and respiratory rate, decrease. It may on occasion be so bad that such a patient is a "nervous wreck" in the morning, but once he has overcome his distress, he becomes quite active, although not overenthusiastic. His sensitivity to weather changes varies. It has erroneously been described by Curry[5] as a reaction of a "cold front type", but without any laboratory foundation.

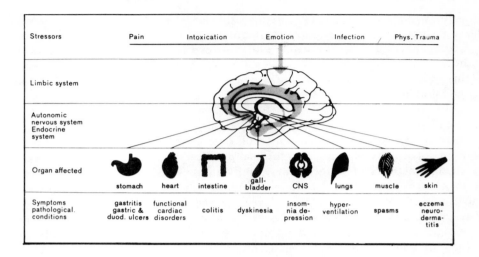

FIGURE 1.   Emotional stimuli acting as "stressors". Stress has been defined by Selye (Montreal) as the impact of external factors on our health. In addition to heat or cold stress, other factors may also impose on our health, notably: pain, intoxication, emotional stimuli, infection, and physical trauma. The so-called Limbic System of our brain reacts to stress via the hypothalamus and pituitary and produces an alarm reaction which stimulates the autonomic nervous system as well as the endocrine system to cope with the stress. The organs affected may be the stomach, heart, intestines, gallbladder, urinary bladder, central nervous system (CNS), lungs, muscles, and skin. Thus many of the pathological symptoms and diseases shown in the last line can be produced by stress.

6. The Sympathotonic Stress Reaction

Sympathotonic people suffer from an overactivity of their sympathetic nervous system. This is due to a tendency to create norepinephrine in their sympathetic nerve endings and norepinephrine and epinephrine in their adrenal gland. Patients are characterized by their euphoria, restlessness, talkativeness, high power of concentration, work power and mental acuity, bodily strength, and aggressiveness. On the negative side, they may suffer from sleeplessness and a speeding up of their vital functions, i.e., blood pressure, pulse rate, and respiratory rate. In women, an additional reaction has been observed, namely dermographia (writing on the skin). This is easily produced by passing a blunt needle over the skin leaving red lines which later turn into long-lasting white lines, due to local production of histamine and norepinephrine — sometimes a delight for lovers.

Additional features in this group are a tendency to heightened emotions, poor appetite which is reflected in their usually slim figure, and "adrenaline constipation" when they become excited by trips away from home. Their reaction to weather changes is very marked and they were described by Curry[5] as a "warm front type".

7. The Serotonin Stress Reaction

The name "serotonin" is derived from its ability to increase blood pressure and create diverse changes of mood and well-being. The serotonin reaction to changes in air electricity, especially air ions, has only recently been recognized in our laboratories.[6] The serotonin patient is particularly sensitive to weather changes and to the positive ion charges of weather fronts. His symptoms are so diverse that their cause can only be verified by urinalysis. Serotonin produces such troubles as sleeplessness, irritability, tension, migraine with nausea or vomiting or visual disturbances, a feeling of electric currents on the skin and hair, swellings, heart palpitations or heart pain, breathlessness (dyspnea), hot flushes with sweating or chills, rhinitis (either hay fever or a stuffy nose), conjunctivitis, sore throat, bronchitis or tracheitis, vertigo, tremor, intestinal hypermotility, and urge to constant urination.[1,2]

A comparison of these complaints with those experienced by the vagotonic and sympathotonic patients immediately shows a wide overlapping. For this reason, a diagnosis of this type of constitution can only be made by urinalysis, as described in the following paragraphs.

8. The Thyroid Stress Reaction

There are people who suffer from occult overactivity of the thyroid gland, a condition not sufficiently recognized in medical practice. This condition is hardly detected by routine methods of endocrinology and we discovered it when we introduced the urinary thyroxin test, in which slight differences in daily thyroid hormone production and excretion can easily be traced.[7] Thyroid people react to all weather fronts — cold, warm, or hot. They cannot stand extremes of heat or cold. They are also very sensitive to changes of air electricity, especially ions and sferics.[8] Their sufferings may be similar to those described in the serotonin constitution but diagnosis can only be based on the fact of increased thyroxin and histamine in their urine. There are, however, typical symptoms which can arouse suspicion of a thyroid constitution: a pulse rate of over 80 per minute; increased metabolism, with a tendency to lose weight; loss of hair; reddening of skin, combined with a strong urge to overactivity which places these patients in the highest category of mental activity. Their superpower is well demonstrated in the high creative output of W. A. Mozart, J. W. Goethe, and A. von Humboldt. Their mental capacity is the driving force of their life, yet on the other hand they may suffer from this because they impose upon themselves goals which are barely attainable.

9. Hormonal Stress Reactions
a. Neurohormones

The effect of air electricity, weather, and climate on neurohormone production and secretion is of decisive importance for the well-being of weather-sensitive people. The following neurohormones are routinely measured by us in the urine in all weather-sensitive patients (Table 1 and Figure 2).

Adrenal hormones — *Epinephrine* (adrenaline) which stimulates our mood, is increased by air electricity and by hot temperatures, yet may become exhausted through protracted tropical heat. *Norepinephrine* (noradrenaline) provides for bodily strength and endurance; it behaves like epinephrine.

17-Ketosteroids (17-KS)* — Provide us with the stamina to cope with environmental stress; they are increased by heat or environmental stress, yet quickly decrease to become converted into 17-OH.

17-Hydroxycorticosteroids (17-OH)** — Represent the cortisone activity of the body; they are responsible for euphoria, sugar, and mineral metabolism during heat stress, but also prevent allergic reactions.

b. Irritation Hormones

Serotonin (5-HT) — Produced in the chromaffin cells of the intestines and in the brain where it serves as neurotransmitter of the psychic reactions, especially in the hypothalamus. Its release from blood platelets by air ions or heat stress increases blood pressure and evokes many irritative reactions, including migraine, as described in Table 2.

---

*   Urinary 17-KS represent one third testosterone and two thirds adrenosterone.
** Urinary 17-OH represent the glucocorticosteroids of the adrenal cortex.

Table 1

URINALYSIS OF NEUROHORMONE PROFILE FOR DETERMINATION
OF CAUSES OF VEGETATIVE DYSTONIA IN WEATHER-SENSITIVE
PATIENTS

No.	Parameters studied	Units per 24 hr	Normal days	Weather front days	Hot days
1	17-KS female	mg	8 — 12	10 — 18	6 — 8
	male	mg	12 — 18	15 — 20	10 — 14
2	17-OH female	mg	2 — 3	3 — 3.8	2 — 3
	male	mg	2 — 4	3 — 5	2 — 4
3	Adrenaline	μg	1 — 4	0.1— 1.6	0.5— 2
4	Noradrenaline	μg	10 — 50	8 — 35	5 — 20
5	Serotonin	μg	0 — 40	50 — 90	10 — 50
6	5-HIAA	mg	2 — 6	5 — 12.5	5 — 9
7	Histamine female	μg	35 — 90	50 — 120	60 — 150
	male	μg	30 — 52	45 — 75	50 — 130
8	Thyroxine	μg	10 — 20	15 — 28	28 — 37
9	Na$^+$	mEq	60 — 110	80 — 140	50 — 125
10	K$^+$	mEq	20 — 32	50 — 40	20 — 35
11	Creatinine	g	1.5— 2.5	2.5— 3.5	2.5— 4.0
12	Diuresis female	mℓ	700 —1500	1000 —2000	900 —1800
	male	mℓ	850 —2000	1200 —2200	1000 —2000

Note: Comparison of excretions on normal days, weather front days, and hot days, showing average
and trend of changes.

Histamine — Released from three types of tissue and blood cells: mast cells, basophil
cells, and blood platelets. Its irritating effect produces allergic reactions in heat-sensi-
tive people. Yet there exist also cold-sensitive people who cannot tolerate a cool
shower, because their histamine release provokes a very unpleasant rash (urticaria) on
the whole body.

Thyroxine — Another interesting observation on histamine release is its combination
with high *thyroxine* release especially on days of weather fronts with high electrical
charges of ions and sferics. This is due to increased metabolism and histidine breakdown initiated by thyroxine.[9]

Acetylcholine and gammabutyric acid (GABA) — These are important neurohormones. They regulate muscular action, yet their reaction seems not to be affected by
climate changes.

c. Sex Hormones

Female sex hormones (estradiol, estriol, and estrone) — Increased by air electricity
and ambient heat. Their release in young animals as well as in children occurs earlier
in the tropical climate than under normal conditions of temperature; menarche in hot
climates begins at 12 years. On the other hand, cold temperatures delay the onset of
the menarche from 13 years in the temperate zones to 14 years in the arctic zones.

Male sex hormones (testosterone and adrenosterone) — Also increased by heat and
especially by ultraviolet sun radiation. This fact accounts for the earlier sexual maturity
of boys and girls in hot climates (12 years) and for a later onset in the moderate climates (13 years) or in the arctic zones (14 years). There has been much speculation on
other reasons for these differences in maturation; they have, however, been disproved
by our controlled animal experiments.[10]

Prolactin — The milk-producing hormone is also affected by heat, not, however,
by normal climatic changes. Its release is low at 4°C and high at 40°C.

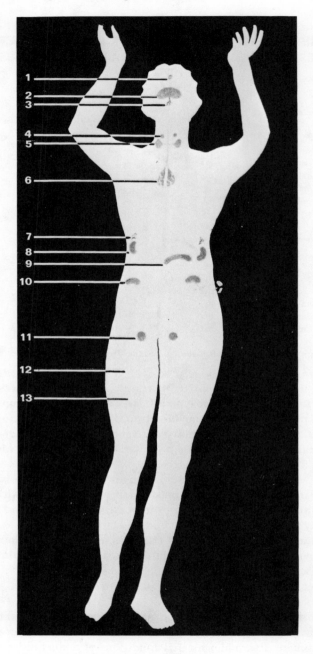

FIGURE 2.   Sites producing hormones and neurohormones in the human body, starting from the top. (1) Pineal gland regulates hormone secretion, contains serotonin and melatonin; (2) hypothalamus controls hormonal secretion and psychic reactions; (3) hypophysis (pituitary) regulates hormone secretion in response to hypothalamus and the pineal gland; (4) parathyroid glands regulate calcium metabolism; (5) thyroid gland regulates general metabolism and mental activity; (6) thymus controls immunity reactions; (7) adrenal gland produces adrenaline and noradrenaline in its core (medulla) and 30 corticosteroids in its shell (cortex). The metabolites of the corticosteroids are mainly 17-Ketosteroids = 17-KS (stress hormones), 17-Hydroxyster-oids = 17-OH (mineral hormones), and male sex hormones; (8) kidney produces urine and contains an anemia-preventing and curing hormone erythropoetin; (9) pancreas regulates sugar metabolism by insulin and glucagen; (10) ovaries produce ovules and female sex hormones; (11) testes produce semen and male sex hormones; (12) mast cells — occurring all over the body — release histamine; (13) chromaffin cells occur mainly in the intestines; they release serotonin, however serotonin can also be produced all over the body and the brain; it is a nervous transmitter which can also induce allergic reactions.

Table 2
THREE TYPES OF REACTION TO HEAT STRESS IN WEATHER-SENSITIVE PATIENTS

Syndrome	Symptoms
Serotonin hyperproduction: Irritation syndrome 415 patients Urinary serotonin and 5-HIAA increased	Sleeplessness, irritability, tension, anorexia, electrified hair, migraine, nausea, vomiting, scotoma, amblyopia, tinnitus, edemata, rheumatic pain in scars, muscles, and joints, palpitations, precordial pain, dyspnoe, flushes with sweat or chills, vasomotor rhinitis, conjunctivitis, laryngitis, tracheitis, vertigo, tremor, hyperperistalsis, polyuria, pollakisuria, photophobia, olfactophobia, audiophobia
Adrenal deficiency: Exhaustion syndrome 420 patients Urinary adrenaline, noradrenaline, 17-KS, 17-OH decreased	Hypotension, fatigue, apathy, exhaustion, "blackout", depression, confusion, ataxia, adynamia, psychic impediment to performing tasks, hypoglycemic spells
Latent hyperthyroidism: "Forme Fruste" 100 patients Urinary thyroxine and histamine increased, serotonin sometimes increased	Insomnia, irritability, tension, palpitations, precordial pain, sweat, tremor, abdominal pain, diarrhea, polyuria, pollakisuria, allergic reactions, reddening of skin, alopecia, orexia with weight loss, overactivity, fatigue, exhaustion, depression, adynamia, confusion

d. Metabolic Hormones

Thyroid hormones — Very sensitive to temperature changes and psychic shock. Their effect on our well-being is decisive since they regulate all our metabolic processes. Not only the metabolism of food but also that of our hormones, enzymes, and brain reactions are affected. The thyroid personality is quick in all his reactions and conceptions, but he is also very sensitive to external influences such as heat or cold.

Growth hormone — Very sensitive to variations of temperature. At low temperature it is increased, as evidenced by the tall stature of the northern population, and at high temperatures it is impaired — no wonder that the pygmies are found in the hot zone. An interesting fact is that nowadays the children in Israel attain a much higher stature than their parents and ancestors. This has been ascribed to their living under controlled cool temperatures; naturally the factor of improved nutrition cannot be excluded.

Pancreatic hormones — Such as *insulin* and *glucagon* are sensitive to cold and heat. This explains the increased food intake in the cold and the lowered appetite under conditions of heat. Here, too, the metabolic effect of the thyroid system is involved. It has been shown that the thyroid-stimulating hormone is increased by low temperatures and suppressed by the higher ones.

Calcium metabolic hormones — The *parathyroid hormone* which increases calcium metabolism and *thyrocalcitonin* which lowers blood calcium. Neither is very susceptible to climate or temperature changes. There exists, however, a third calcium hormone, the well-known *vitamin D (calciferol)* which is produced in the skin by ultraviolet sun radiation and mobilizes calcium. This effect is the secret of the prevention of infantile rickets by sun radiation. Yet it has its drawback in adults: calcium mobilization carries the calcium into the urine and may produce kidney stones. This would explain the high incidence of urinary calculi in people moving from the moderate zone into a sunny climate. Moreover, chronic sun radiation may produce skin cancer.

e. Posterior Pituitary Hormones

There are three hormones in the posterior lobe of the pituitary, all of which react to our surroundings. The *melanophore hormone (MSH)* is released by sun irradiation and contributes to the tanning reaction of the skin. The *uterus-contracting hormone (oxytocin)* is released by darkness and seems to be responsible for the increased birth rate at night time. The *antidiuretic hormone (ADH, vasopressin)* regulates our diuresis. None of the hormones are affected by climate changes.

10. Timing of Stress Reactions

A clinical study combined with urinalysis of some 100 weather-sensitive patients has shown that there exist three different syndromes of acute reactions to ambient heat changes in air electricity, especially ions and sferics (Table 2).

The prethermal syndrome (serotonin hyperproduction, irritation syndrome) — Irritability, tension; migraine with nausea, vomiting, amblyopia or scotoma; edemata, palpitations, precordial pain, dyspnea, hot flushes with sweating or chills, ''allergic'' complaints such as vasomotor rhinitis, conjunctivitis, laryngitis, pharyngitis, tracheitis, hyperperistalsis or diarrhea, polyuria or pollakiuria; scar ache, rheumatic pain, anorexia, tinnitus, tremor, and vertigo. The most typical characteristic of this group is that people start suffering before the arrival of the weather fronts. This reaction is due to hypothalamic serotonin output provoked by a steep increase in positive ionization of the air which the weather front pushes ahead before its arrival. Patients suffering from the serotonin syndrome are amenable to prophylactic measures, such as using negative ionizers, and to drug therapy with specific serotonin antagonists.

The midthermal syndrome (intermittent hyperthyreosis) — Hyperthyroidism (Forme Fruste) with few clinical symptoms, but typical tachycardia. Urinary neurohormone excretion is generally increased. The increase of urinary histamine excretion (above 90 μg/day) and urinary thyroxine (above 20 μg/day) is pathognomonic. This form of slight hyperthyroidism is treated with beta-blockers and lithium (rarely with propylthiouracil or methimazole).

The postthermal syndrome (adrenal exhaustion syndrome) — Hypotension, apathy, depression, fatigue, lack of concentration or confusion, exhaustion, hypoglycemic spells, and ataxia. The most typical complaint of this group is that they suffer from climatic heat stress progressively more each year and and urinary catecholamine excretion becomes particularly low. Drug treatment of this condition is possible by minidoses of monoamine oxidase blockers to compensate for catecholamine deficiency.

The effect of all treatments can be controlled by urinalysis. Naturally, adipositas and hypertension have also been recognized as conducive to heat complaints.

B. Psychiatry
1. Problems

It is hard to separate the pyschiatric reactions in weather-sensitive people into groups due to electric changes and groups due to climatic changes in temperature or humidity.

Surely, sufferings from mental diseases are influenced by weather changes, as Faust[11] has shown for Switzerland, where his statistics could classify hospital admissions during the following four weather conditions: foehn, cold fronts, occlusion, and warm fronts. The following results emerged.

2. Diseases

Schizophrenics (especially paranoic and hebephrenic schizophrenics) react to foehn, occlusion, and warm fronts. Catatonic schizophrenics are irritated by incipient cold fronts; cases of schizophrenia simplex seem to react to occlusions. Schizophrenics tend to overcome weather reactions quickly and easily because of their labile neurohormone reactions.

Depressives are particularly sensitive to weather conditions. They can sense electric weather changes by increased serotonin production. Sufferers from periodic depressions combined with exhaustion are sensitive to warm fronts; they are least affected. Reactive or neurotic depressives and depressives of involution as well as other depressive types react to foehn, cold fronts, occlusion, or warm fronts; they are the most affected.

For lack of data it is difficult to get objective findings of *neurotics;* it seems that crises in puberty and neuroses of character are aggravated by foehn, whereas obsessional and anxiety neurotics suffer more from cold fronts or occlusion. In any case, they are especially liable to react to climatic events.

Psychopaths seem to be affected by neutral air occlusions, *arteriosclerotics* by cold fronts and cold air occlusion, *alcoholics* and *drug addicts* by foehn, cold fronts, neutral air occlusion, and incoming warm fronts.

3. Hospitalization

Clinical experience shows that certain psychiatric patients have to be hospitalized more frequently during certain periods of the year because of their deteriorating condition. The maximum of admissions was found always in summer and sometimes in spring: *schizophrenics* of both sexes most frequently during the warm months (summer, second half of spring, autumn); *endogen depressives* most frequently in spring (men) and summer (women); *neurotics* and *psychopaths, arteriosclerotics, alcoholics,* and drug addicts most frequently in summer and spring.

4. Conclusions

There is no doubt that psychiatric patients tend to suffer from aggravation of their sufferings by weather changes. One should, therefore, take care to increase their dose of medications whenever weather changes may require such a step.

C. Neurology
1. Problems

The main neurologic disturbances in weather-sensitive patients are those connected with insomnia and anxiety, weakness and faintness, headache and migraine, rheumatic pain and scar ache, epilepsy and hysteria. Their mere existence predisposes a patient to the aggravating influence of weather changes. As most of the above complaints may be intensified by serotonin production, we have to reckon with the specific effects of ions and sferics which release serotonin. Additional neurohormones must be taken into account, especially epinephrine, norepinephrine, histamine, and thyroxine.

2. Insomnia and Anxiety

Insomnia and anxiety may be caused by a surplus of serotonin, epinephrine, norepinephrine, histamine, or thyroxine. An intellectual person may pursue his ideas at nighttime, being unable to cut them off due to a surplus of catecholamines. A thyroid person may suffer even in the daytime from the mental and cardiac stress imposed by a surplus of thyroid hormone. In both cases the occurrence may be an occasional one, but when it coincides with weather change and serotonin release, an unbearable situation may arise for the unfortunate patient, who has to learn his drug of choice, be it an hypnotic, a sedative, an antiserotonin drug, an antithyroid drug, or sometimes an antihistaminic.

3. Weakness and Faintness

The underlying cause may be hypertension or hypotension. Both may be aggravated by the exigency of a weather change. Heat, cold, and sometimes changes of atmospheric electricity have been shown by us to precipitate extreme changes in catecholamine secretion, which could upset a weather-sensitive person and precipitate syncope.

The most common complaint of the hypertensive person is dizziness due to weather change. On the other hand, the hypotensive may react by extreme fatigue, lassitude, apathy, exhaustion, lack of concentration, and confusion culminating in depression. The hypotensive patient would react well to treatment with MAO-blockers (see Section III.C).

4. Headache and Migraine

Headache, if due to histamine (cluster headache), is naturally sensitive to weather changes involving air electricity, extreme cold or heat, and dryness. In this case a changing weather front may release additional histamine. The treatment with antihistaminics may be disappointing; for this reason we prefer antamines which counter both histamine and serotonin (see Section III.C). Classical migraine — mostly due to serotonin released by the glomus caroticum of the typical migrainous patient — is very sensitive to weather changes.[12] In Israel the sharav is the most "reliable" provocation of a migraine attack. Our preventive treatment with antiserotonin drugs is therefore instituted during the sharav seasons (spring and fall) and discontinued in summer and winter.

5. Rheumatic Pain and Scar Ache

Rheumatic complaints and scar ache are hard to define. They harass people especially sensitive to weather changes due to ions and sferics by serotonin release. As these electrical changes arrive 1 to 2 days before the atmospheric weather changes and release serotonin, it is the rule that the rheumatic and "scarred" patients claim that they can prophesy weather changes. Unfortunately, the inexperienced physician tends to send them to a psychiatrist, yet the appropriate therapy would be to give them an antiserotonin drug or to keep them under an ionizing apparatus combined with an AC/DC electrical field.

6. Epilepsy and Hysteria

Epileptic reactions are sometimes difficult to separate from hysteria; the electroencephalogram (EEG) gives the right answer. As a more subjective guide, an epileptic fit is usually combined with urinary incontinence, whereas the hysteric patient would take care to avoid this offensive and compromising reaction. Both disorders react heavily to electric changes such as ions and sferics. Protection of the epileptic patients would require an increase of their daily intake of antiepileptic drugs, whereas the hysteric persons would need a sedative antiserotonin drug. Their reaction toward their children gives rise to concern on days of electrical air changes, as it may seriously affect the "battered child".

D. Psychosomatic and Psychotechnic Problems
1. Psychosomatic Complaints

Psychosomatic diseases may encompass every organ of the human body. They reflect the impact of an organic disease on our mental reaction. It is, therefore, impossible to provide a description of psychosomatic complaints that may be aggravated by weather influence. One example needs consideration: bronchial asthma. Bronchial asthma is an organic disease that tends to acquire a secondary emotional aspect, since the patient normally tries everything to avoid an attack. As this wish cannot always be fulfilled due to the interference of organic reactions, such as infection, allergy, exposure, or stress, it is only natural that every weather change would be able to elicit an asthma attack. In such cases, the treatment can only be symptomatic. We have tried ionizing treatment and enzyme induction by proxibarbal — both with ambiguous results, since the emotional factor cannot be excluded.

2. Psychotechnic Problems

The impact of weather change on hormonal disturbances was also reflected in psychological test performances during climatic heat stress from sharav winds carried out by Rim[13] in Haifa (Israel). Candidates for office posts or technological jobs were subjected on normal days and on sharav days to three series of psychological tests. On sharav days the results were as follows:

Intelligence tests — Impaired reactions in domino test, Miltha verbal intelligence grouping, understanding written instructions, practical mechanical comprehension.

Performance tests — Unchanged abilities in clinical skill, arithmetic performance, substitution (numbers for letters), spatial perception of two-dimensional geometric shapes.

Neuroticism tests — Higher scores in Maudsley personality invention and Eysenck's psychoticism scale.

The performance of the first group was improved by apparatuses providing negative air ionization; the second and third groups did not respond favorably.[14]

Charry[15] tested the effect of positive air ionization in the following way. In New York City, 86 subjects were exposed to both positively ionized and nonionized environments over two different experimental sessions in a counterbalanced, within-subjects repeated, cross-over design. The positively ionized atmosphere ranged from 20,000 (+) ions per cubic centimeter to 30,000 (+) ions per cubic centimeter and on nonionized days the background ranged on the average from 0 to 300 (+) or (−) ions per cubic centimeter. Measures of performance (complex and simple reaction time tasks), mood (affective state as assessed by the Adjective Check List and Questionnaire), and physiological activation (skin conductance) were taken for both days. The results were analyzed on the basis of two different measures of ion-sensitivity; Lacey's ALS (autonomic lability score) and Sulman's PSSI (physical sensitivity to serotonin irritation) measure.

Sensitive subjects behaved in significantly different ways under ionized (vs. nonionized) conditions — in psychological state (decreased urgency), physical state (increased fatigue and organic symptoms), depressed physiological activation (prestress or stress), and for performance (increased simple reaction time).

When age, sensitivity, and ion condition were considered simultaneously, two significant three-way interactions emerged for tension and simple reaction time. For simple reaction time, it was the younger sensitive subjects who were most influenced by the *positively* ionized air, demonstrating increased reaction time compared to their performance in the nonionized condition. For the measure of tension, older sensitive subjects exhibited significantly higher levels in the ionized compared to the nonionized condition.

On the basis of this study, it would appear that significant changes in affective, cognitive, and physiological states are experienced by subjects under high *positive* ion conditions due to weather change (Table 3).

3. Electroencephalogram

Our group[16] has studied the effect of *negative* air ionization in the human EEG. The results were as follows. The EEG, after negative ionization, shows an advance of the normal alpha rhythm from the occipital brain to the forebrain which promotes the conception of ideas; a stabilization of the frequency at 10 Hz, indicating relaxation, and an increase of amplitudes which improves work capacity. Last but not least, there occurs a synchronization of the right and left brain hemispheres, which means a balancing of the personality (Figure 3). When, however, we tested negative air ionization in students who underwent the above-mentioned standard tests of psychotechnics, no significant improvements of performance were noted.[17] Reaction can be shortened, however, in simulated car driving tests.[18-20]

Table 3
REPORTED PHYSIOLOGICAL
EFFECTS OF POSITIVE IONS

Inhibition of growth of tissue cell cultures
Increased respiratory rate
Increased basal metabolism
Increased blood pressure
Produced headache, fatigue, nausea
Produced nasal obstruction, sore throat, dizziness
Increased skin temperature
Depressed rate of ciliary activity
Increased muscle chronaxie
Altered alpha wave of the EEG
Reduced succinic oxidase activity in the adrenals
Increased the blood level of 5-hydroxytryptamine

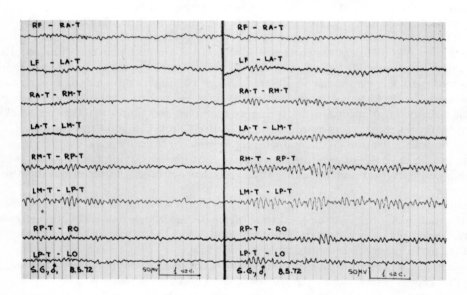

FIGURE 3.   Negative air ionization has a specific effect on the brain as shown in the electro-encephalogram (EEG). The different leads taken right (R) and left (L) from the frontal area (F), anterior (A) and posterior (P), temporal area (T), as well as from the occipital area (O) show the following improvement in brain reactions. (1) Frequency stabilized from 8 to 12 Hz at 10 Hz allowing relaxation; (2) amplitudes increased by 120% promoting better thinking; (3) brain waves spreading from the perceptive occipital to the frontal area, creating initiative and spontaneous conception; (4) synchronization of the reaction between right and left brain hemispheres, balancing the personality.

A survey of the vast literature on the beneficial effects of and improved performance in man under negative ionization has been published by one of our patients.[21] (See also Table 4.)

Psychotechnic experiments with negative air ionization were carried out on laboratory animals by us and by many other authors. A perusal of the extensive literature[22] shows that in acute experiments all factors of performance, endurance, learning, and survival were improved; however, in chronic experiments these benefits disappeared. In female mice and rats, the estrus cycle was speeded up[23] and in pregnant rats subjected to serotonin abortion, the detrimental effect could be abolished by negative ionization.[24]

Table 4
REPORTED PHYSIOLOGICAL
EFFECTS OF NEGATIVE IONS

Decreased respiratory rate
Decreased basal metabolic rate
Decreased blood pressure
Produced a feeling of well-being
Increased vital capacity
Decreased skin temperature
Acceleration of the conversion of succinate to
  fumerate
Stimulation of cytochrome oxidase activity
Decreased eosinophilia and lymphocyte count
Increased CO_2 combining power of plasma
Decreasd blood sedimentation rate
Decreased muscle chronaxie
Increased ciliary activity
Increased frequency of mitosis
Increased resistance to infection
Suggested as therapy in chronic rhinitis, sinusitis,
  migraine, insomnia, tuberculosis, wound and burn
  healing, asthma, hay fever, emphysema,
  bronchitis, conjunctivitis, chlorine gas poisoning

4. Psychotechnic Studies of Electrofields

Lang and Lehmair[25] applied the Elevit G & I®* Electrofield Apparatus in school-rooms. The apparatus produces electromagnetic impulses from a DC field of 4000 V (G for Gleichstrom) combined with 10 Hz rectangle impulses (I for Impulse) of variable amplitudes (normally 20 V) and frequency (normally 100 msec). The electrode elements vary according to the size of the room between 0.4 and 2 m². The larger size is sufficient for a room of $5 \times 6 = 30$ m². The positive electrostatic DC field spreads from the ceiling to the negative field of the floor (Figure 3). The exposure of 54 pupils to the apparatus showed 25 to 30% improved observation capacity, enhanced work perform-ance, perfected conception and perception, and increased immunity to influenza epi-demics.

Koenig and Lang[26] applied the Elevit apparatus to 48 probands in a car driving simulator, using only 1000 V DC, instead of 4000 V. The apparatus did not shorten reaction time; however, it decreased the occurrence of driving mistakes by 10% and improved the management of crash situations in labile probands. Repeating the exper-iment in 300 drivers yielded similar results.

E. Crime

Homicide has been studied in Israel by Landau and Drapkin.[27] Their statistical eval-uation confirmed the general experience that during the months of extreme sharav heat (March-April-May and August-September) acts of aggressiveness are at least twice as prevalent as in cooler months. Still, June and July are also hot months in Israel. This shows that it is not the heat effect alone but the mental instability created by high air electricity serotonin secretion during sharav periods that must be held responsible for aggressiveness. Similar observations have been reported for the moderate zone by Faust.[11] The following symptoms are disturbing and frequently produced by stress due to changing weather: weariness, indisposition, multiple mistakes, epileptical seizures, listlessness at work, headaches, respiratory distress, troubled sleep, bad temper, nerv-

* Baldham, Munich, West Germany.

ousness, increased forgetfulness, and head pressure. It is striking that drug addicts show a more varied picture than depressive patients. The correlation of increased sensibility (pain in old bone fractures or operation scars) with aggressiveness remains open. The seasonal increase of symptoms due to meteorological influence with peaks in spring and autumn falls in a changeable weather period (i.e., a time when intensive electrical changes are expected to harass the electrosensitive patient). A specific report on crime under the influence of Santa Ana winds can be found in Section II.A.

F. Suicide

The influence of weather and weather fronts on suicidal traits has been described in up to 5000 publications. Faust[11] has recently compiled statistics in this field. His results indicate clearly that the depressive patient (a high suicide risk) is especially prone to commit suicide when a weather front with its electrical changes comes up. In addition, it is the fair foehn weather that brings the depressive patient to the decision that he cannot cope with the amenities and enjoyments of fair weather. This may induce him to commit suicide, and again the electrical changes appearing with the foehn condition should be incriminated. Follow-up examinations of epinephrine and norepinephrine in weather-sensitive depressive patients have shown us that they are prone to suicide when their catecholamine values are not too low, but at improving low middle values (i.e., high enough to allow them a certain degree of initiative). In addition, it emerges that the suicide victim is *a priori* psychically disturbed. Of depressives, neurotics, psychopaths, alcoholics, and other addicts, 50% are sensitive to weather changes and therefore also prone to suicide. In the group of schizophrenic disturbances, the percentage of suicide tendency is much lower, since these patients live in a world of their own. One can, however, conclude from a number of surveys that actually 39 to 50% of all suicide cases are objectively affected by weather conditions, especially the fair weather of spring. A survey of suicides in Israel has shown an insignificant peak during the months when sharav heat prevails (March-April-May and August-September). Yet, for Hamburg, Kuhnke[28] showed a significant increase in suicide on days with southwestern warm air influx with high counts of sferics.

An inquiry was held in Munich into the number of suicides during a foehn. This had been done before, but without giving enough weight to the effect of weather conditions; the results were not reliable. The new inquiry established that the number of suicides rose during a foehn with the arrival of the warm air and the delayed departure of the front. This clearly indicates that, apart from individual reactions, ions and sferics appearing with changing weather conditions can affect anybody who is prone to suicide.[11]

G. Accidents and Mortality

Statistics show that mortality is high in winter and low in summer. This is especially due to the high mortality from angina pectoris, coronary thrombosis, and arteriosclerotic diseases under cold stress. In summer, hot spells increase mortality only after the first day of heat, when the adaptation mechanisms fail in diseased or elderly people. The same holds for periods of extreme heat fluctuations on weather-front days with high air electricity.

It is generally recognized that the factors tending to increase traffic accidents are extreme sunshine and high temperatures; while rain, snow, fog, and low temperature call for vigilance and reduce traffic incidents. Weekend traffic does not change these rules. Our neurohormone studies have shown that extreme heat produces the following changes conducive to traffic accidents:

1.   Low ketosteroids which impair coping with stress situations

2.   Reduced epinephrine levels which lessen concentration
3.   Reduced norepinephrine levels which minimize body reactions
4.   Increased serotonin and 5-HIAA levels which create impatience and reduce consideration of impending danger
5.   High histamine and thyroxine levels which impair our sound judgment

Factors 4 and 5 are due to increased air electricity, especially ions and sferics.

It is obvious that the same rules hold also for work accidents, where an additional factor may induce accident proneness: high sweat rate induces sodium loss compensated in extreme dehydration by flooding of the blood with potassium. As potassium reduces the reactivity of all muscles, potassemia prevents quick muscle action required for controlling dangerous situations created by sewing machines, electricity breakdowns, and car braking.

The above neurohormone changes being connected with the increase of air ions and sferics would also explain the higher incidence of heart mortality and cardiac insufficiency conducive to heart infarction on days of extreme heat stress, as Santa Ana, foehn, and sharav winds.

H. Thromboembolism

Thromboembolism may appear in man like an epidemic on special days. It affects patients after operations especially in urologic departments, or those suffering from varicose veins, phlebectasias, heart infarcts, cardiac insufficiency, arteriectasias, arterial occlusion, and polycythemia vera. Its correlation to weather changes has often been described. Our research has shown that air ionization indeed provokes neurohormonal changes, especially serotonin release, which may precipitate thromboembolism. Recently we have extended these studies to sferics showing that they affect serotonin release as does air ionization.[29] Thus, it appears that the "epidemic" occurrence of thromboembolism on days of high positive electric air charges is a phenomenon that can be avoided. Its combating by suitable air neutralization is presently being studied by us;[30] negative electrical fields and negative ionization may provide the cure. The problem found wide interest in Germany when Rehn[31] published his observations on the abolition of thromboembolism in an area (Kahlenberg) rich in negative ions and the return of thromboembolism in his hospital when the air was deprived of its ionization (Kahlenberg factor). Recently, we showed that negative air ionization can prevent postoperative thromboembolism.[32]

A new method for avoiding postoperative thromboembolism has therefore been introduced by us, employing negative air ionization (Figure 4). In a first trial, patients prone to thromboembolism were kept in a room under permanent negative air ionization. Groups of four patients each stayed in two rooms of 5 × 5 m size in which four ionizing apparatuses each had been installed. The patients remained there for an average period of 10 days after urologic operations; 228 patients passed through these rooms during 28 months. Every patient lay at a distance of 2 m from a "Modulion"® ionizer emitting an average of 1×10^4 negative ions per cubic centimeter of air. In the 6 control rooms, 1232 postoperative cases were hospitalized without ionization. Other treatment in all rooms was identical, avoiding anticoagulants.

In the 228 postoperative patients exposed to permanent negative air ionization, there occurred only one case of thromboembolism during an observation period of 28 months (0.43%), whereas in the 1232 patients in the 6 control rooms 12 (0.97%) suffered from thromboembolism and had then to undergo standard treatment with anticoagulants; 3 of them died. The percentage of postoperative thromboembolism (0.97%) corresponded well to the average number of thromboembolism encountered in the rest of the hospital and to other hospitals in the country which did not use air

FIGURE 4.   Modulion® ionizer without fan (Amron — Amcor, Tel-Aviv). Four needles emit negative ions. It can provide power supply for up to 18 needles.

ionization. Thus, it appears that negative air ionization can replace the risky prophylactic use of anticoagulants after operations.

I. Working Efficiency and Acclimatization

While men who are accustomed to working in the heat should be instructed in the methods of hot climate hygiene, the problems of personnel management are unquestionably most acute when unacclimatized men are exposed to heat or extreme charges of air ions and sferics. Men will work efficiently only if they are fully acclimatized to the existing conditions. This acclimatization includes mainly reduction of their sweat loss, i.e., loss of sodium. Unacclimatized men cannot be expected to be fully productive during their first 2 weeks of exposure. To obtain maximum productivity from men during their early weeks of exposure to heat and to develop acclimatization as economically as possible, they should be employed in cool surroundings during the morning and in the heat in the afternoon; men who work continuously during these afternoon exposures will develop the physiological adaptations to heat as rapidly as they will from whole day exposures, It is of greater practical benefit if the men work for only 2 hr at the rate that will later be required of them, than to allow them to stay in the heat for longer periods at reduced rates of work. If the conditions of heat exposure are severe, it may be necessary to adopt a two stage acclimatization procedure such as has been utilized satisfactorily in South African gold mines.

When men who are not accustomed to heat attempt to perform moderate work in a hot environment, they may overheat within 1 to 2 hr, showing markedly elevated rectal and skin temperatures and reduced gradients between internal and surface temperatures, metabolism increases in proportion to body temperature; heart rate is nearly maximal, and the subjects show signs of circulatory instability such as syncope, which is most pronounced when they attempt to stand. Exhaustion is imminent. With repeated daily exposure to the same combination of work and heat stresses physically fit men acclimatize rapidly — in four to eight daily exposures. Acclimatization is characterized by:

1.   Gradual improvement in evaporative cooling efficiency and in the sensitivity and capacity of the sweat mechanism

2. Gradual improvement of temperature regulation by the 8th day of exposure, enabling men to perform work in heat (if the heat stress is not too intense) with increased rectal-to-skin temperature gradients, but with about the same elevations of core temperature and metabolism as when they perform the same task in a cool environment
3. Markedly improved circulatory stability with reduced heart rate
4. Rapid reduction of renal Na^+ output and a much slower reduction of sweat Na^+ concentration in response to Na^+ deficit
5. Rapid compensatory reduction of renal water output if dehydration occurs

The time course of the remarkable transformation that occurs during heat acclimatization may be affected by:

1. Dehydration in which sweating, evaporative cooling, and circulating plasma volume decrease, increasing circulatory strain and elevating body temperature in the heat
2. Physical fitness of the subject, fit men adapting more quickly and completely to work in heat, and strenuous winter training programs for varsity athletes preacclimatizing them to work in heat
3. Physical activity during the acclimitization period (sedentary men do not acclimatize fully to work in heat)

The most efficient way to acclimatize men for work in heat is by daily 100-min periods of treadmill work at room temperature.

The question of whether internal body temperature of resting men is permanently elevated in hot climates was debated for many years. However, it has been found that it probably is not, either in long-term residents of hot climates or in subjects during experimental studies of heat acclimatization. Basal metabolism of healthy individuals increases during the first week of exposure to a hot climate, then falls gradually until after several months of residence in the hot climate, it is below ordinary basal standards. There is some evidence that the basal metabolism of men decreases during residence in hot climates and that this occurs in natives as well as in Europeans who have moved to the tropics. Numerous investigators have found evidence of reduced metabolism, ranging from 6 to 24% in residents of hot climates.

Acclimatization will be retarded if there is a deficit in water, sugar, or salt balance, and it is therefore particularly important during the period of acclimatization to ensure ample water, sugar, and salt intakes. In addition, acclimatization is developed faster by men who are in good physical condition, and it will be beneficial to ensure that men are properly trained in their work before they are exposed to heat and air electricity and thereafter that they remain physically fit. Men who have been off work for any reason, for periods of between a few days and a few weeks, ought to be regarded as only partially acclimatized; present evidence suggests that even after 2 to 3 weeks absence, reacclimatization is complete in a few days. Personnel who are absent from work for longer than about 3 weeks should be treated as unacclimatized.

Men ought not to be employed in hot surroundings if they are (1) obese; (2) prone to cardiovascular disease; (3) suffering from skin diseases; (4) recuperating from febrile illness; (5) over 45 years old, unless there is good evidence of an individual's ability to work in heat without distress.

All clothing worn in hot climates should be made of lightweight material and ought to be loosely fitting, so as to interfere with the evaporation of sweat as little as possible and to allow horizontal as well as vertical ventilation.

J. Recreation

Few papers have been published on the influence of weather on outdoor recreation. Clawson[33] points out that "solid professional discussion of outdoor recreation is always difficult," since nearly everyone has experienced outdoor recreation personally, and he naturally feels he knows everything about it. Even a modest acquaintance with outdoor recreation suggests, however, that it is weather prone, and anyone who has had a picnic spoiled by a sudden downpour can testify to that. Nevertheless, Clawson suggests that we need to explore in just what ways and at what stages in the recreation experience, weather and climate are most influenced.

Clawson[33] goes on to describe some basic considerations about the nature of outdoor recreation in order to better understand the basic relationships that weather is likely to influence. These include five interrelated sequential stages: (1) anticipation or planning, when the person or family considers what to do, where and when to go, what to take, and how much time and money to spend; (2) travel from home to the site; (3) on-site experiences; (4) travel home again; and (5) recollection, when one relives the earlier experiences, and perhaps reinterprets them differently than one did at the time, based on the many photos taken, when they are assembled in an album.

Most forms of outdoor recreation are dependent upon a certain range of temperature, sunshine, humidity, wind velocity, and other climatic factors, if they are to be enjoyed. Climate also affects the kind of environment in which outdoor recreation takes place, the most obvious effects being upon water supply, vegetation, and snow for skiing. In addition, the fact that leisure or free time for outdoor recreation is rather rigidly controlled by time is important, in that one must go when one has free time, not when weather is favorable. Moreover, one could suggest that the elements of available time and money form a ratio that in recreation terms may be termed "desirability".

The possible value to recreation of weather modification may be due to climate changes, but it is clear that more accurate and longer-range weather forecasts would also have considerable effect. Perhaps, therefore, we could do well to wish for the ideal recreational climate, viewed by Clawson as one where it never rained, was always pleasantly warm but not too hot, was always mildly sunny, was never too humid, had only gentle breezes, etc., and yet it would have luxuriant forest and other vegetation, flowing clear pure streams, cold enough for trout but warm enough for swimming, and possessing various other desirable attributes.

The first step in this direction was taken by Harlfinger.[34] He published a vacation guide on the bioclimatology of the Mediterranean countries including the European coast of the Atlantic and the Black Sea. It contains all vacation sites and details of their climate.

K. Sport and Exercise

Jokl[35] has described the syndrome of *effort sickness.* "Effort sickness" establishes itself most commonly after brief exhaustive performances in running, rowing, tennis, football, wrestling, swimming, and other branches of athletics. It is characterized by headache, nausea, vomiting, and a pronounced feeling of weakness but does not interfere with consciousness. Similar symptoms have been described as "indisposition after running", "mountain sickness", "air sickness", "sea sickness", or "athlete's sickness".

Some athletes have their first experience with "effort sickness" when they compete during the early stages of a previously unrecognized infectious disease, such as influenza, measles, gastrointestinal upsets, or acute poliomyelitis. Save in such exceptional circumstances, "effort sickness" does not cause permanent physical harm. Its prognosis is good and subsequent athletic performances are not affected by it. Jokl once

Table 5
URINARY NEUROHORMONE
LEVELS OF RUNNERS

Stress hormones decreased
    17-KS from 12.0—8.0 mg/l
    17-OHS from 3.8—2.5 mg/l
    5-HIAA from 5.4—3.5 mg/l
Fight and flight hormones increased
    Epinephrine from 0.7—3.6 μ/l
    Norepinephrine from 22—41 μ/l
Metabolic hormones increased
    T-4 (urinary) from 10—15 μ/l
    Histamine from 30—42 μ/l

Note: All changes were highly significant (p <0.005).

suffered three times on the same afternoon from "effort sickness", after the heat, the semifinals, and the final of a 400-m race in which he ran his best time of the season.

Symptomatologically, "effort sickness" is indistinguishable from "acute mountain sickness" with its low oxygen tension and high concentration of air ions and sferics. Mild physical exertion at high altitudes causes attacks as severe as violent exertion does at sea level. Signs and symptoms of "mountain sickness" invariably disappear during the descent. The symptoms of physical exhaustion after maximal athletic performance at sea level may be analogous to those experienced by mountaineers. The cause of the bodily disturbance is in both instances the same: lack of oxygen ("anoxia") caused by the increased metabolic requirements brought about by muscular effort, or by the lowered oxygen tension in the atmospheric air, complicated by serotonin release from air ions and sferics.

Hyperventilation due to diminishing oxygen pressure of the atmospheric air at high altitudes also engenders a continuous loss of carbon dioxide; thus it not only induces anoxia but also acapnia, and in turn, intensifies the deficiency of oxygen. The reason is that carbon dioxide is the physiological stimulus for breathing, so its disappearance removes the most important regulatory mediator of respiratory adjustment. This is one of the reasons why acute "mountain sickness" as well as "effort sickness" do not occur immediately following strenuous physical effort, but only after an interval of several minutes. The same mechanism may be operative in subjects who "become sick" before a race or game when psychological factors trigger hyperventilation.

Thus, either high altitude or maximal physical effort may overtax the adaptive resources of the healthy but untrained human organism. Negative air ionization seems to be able to overcome these drawbacks. In Russia, extensive experiments of "doping" by negative ionization have been carried out by Minkh.[36] Good results have also been reported by Rivolier et al.[37] For the time being, this type of doping has not been objected to by the Olympic Committee, and it may provide the explanation for the excellent performances of the Russian contingents at the recent Olympic games.

Sulman et al.[38] studied the effect of the weather on sport performance. Typical results were obtained in long-distance runners. Thirty male students 20 to 25 years old, training for a long-distance contest, had their urinary neurohormone levels tested before and after running 5000 m at an ambient temperature of 10 to 15°C. The runners to reach the tape exhibited the typical average pattern of reaction, as shown in Table 5.

The results are compiled in Table 6. Extreme weather conditions (such as heat or cold) or disease of the probands (such as influenza) reversed the typical pattern completely.

Table 6
MOOTED MECHANISMS OF NEUROHORMONE REACTIONS IN LONG-DISTANCE RUNNERS

Hormone	Function	Reaction	Explanation
17-KS	Androgen metabolite	Decrease	Physical ability to cope with stress
17-OH	Cortisone metabolite	Decrease	Mobilization of sugar and sodium
5-HIAA	Serotonin metabolite	Decrease	Mental ability to cope with stress
Norepinephrine	Fight hormone	Increase	Improves circulation and mobilizes sugar
Epinephrine	Flight hormone	Increase	Improves heart action circulation and sugar metabolism
Thyroxine ⎱ Histamine ⎰	Metabolic hormones	Increase	Required for increased metabolism and heat production

L. Military Activity

Fatigue of soldiers in the tropics only differs from fatigue elsewhere in the world because it is aggravated by a warm and humid climate and by serotonin release due to high ion or sferics count. When tropical fatigue is severe, there exists usually a basis of hot climate fatigue with a neurotic feature superimposed as a result of nonclimatic stresses inducing adrenal exhaustion. The importance of the neurotic component will vary with the type and degree of psychological stress and the personality of the patient who complains of "tropical fatigue".

The degree of physical deterioration is insufficient to account for the inefficiency and poor morale apparent in all activities. Isolation, boredom, conditions of service, and the moral structure of the men themselves are conducive to drug abuse and addiction, encountered in more than 50% of the service personnel in tropical as well as in temperate zones.

European civilian residents of the tropics have generally chosen to work there and are usually free, within certain socially acceptable limits, to modify their way of life to suit the peculiarities of the climate. Life in the tropics generally presents them with both more opportunities and more stimuli than life in their native country. The climate appears to be regarded by most long-term residents as only one of many potential sources of irritation and as only one of many factors contributing to the "fatigue" reported to develop towards the end of a continuous tour of service, due to adrenal exhaustion.

Military personnel stationed in the tropics have not chosen to be there but have been sent there. Many regard their stay as a temporary hardship to be endured, and, when not on duty, fail to compensate for the lack of the familiar social and recreational facilities of their home environment. During combat activity, military personnel are immensely stimulated by wake hormones, such as amphetamine (Pervitin of Hitler's army), yet it may produce adrenal atrophy and military demoralization. Modern military medicine uses MAO blockers, which slow down the metabolism and breakdown of endogenous epinephrine and norepinephrine (the fight and flight hormones) without affecting the adrenal gland itself or inducing addiction.

Naval personnel living aboard ship in the tropics probably with low ion concentration are in another category. They carry much of their familiar environment with them in the form of the ship itself and life at sea. It is among these groups that the effects of heat are probably most reliably identified. An increased sickness rate, especially of skin diseases, persistent awareness of the heat, increased feeling of lassitude or irritability, and lowered output or efficiency have been reported. All these complaints are an outcome of adrenal cortical and medullary exhaustion combined with serotonin release due to high air electricity.

II. MEDICAL IMPACT OF EVIL WINDS

A. Santa Ana and Similar Winds

1. Definition

Santa Ana is a hot, dry, foehn-like desert wind, generally from the northeast or east, especially in the pass and river valley of Santa Ana, Calif., where it is further modified as a mountain-gap wind (Figure 5). It blows, sometimes with great force, from the deserts to the east of the Sierra Nevada Mountains and may carry a large amount of dust. It is most frequent in winter; when it comes in spring, however, it does great damage to fruit trees. In fall, winter, and the early spring months, occasional foehn-like descending (Santa Ana) winds come from the northeast over ridges and through passes in the coastal mountains. These Santa Ana winds may pick up additional amounts of dust and reach speeds of 50 to 75 km/hr in north and east sections of Los Angeles, with higher speeds in outlying areas to the north and east. They rarely reach coastal portions of the city because they are broken there by the western sea breeze.

The southwest portion of Los Angeles between Long Beach and Santa Monica is seldom affected by Santa Ana winds, being distant from mountain pass influences. The northwest winds, however, are strongest over open level country, and reach their highest velocities over the coastal plain between Santa Monica and Oxnard. One of the most deceiving weather patterns near the coastal area is the Santa Ana wind. A Santa Ana wind may change a flat blue sea into an area of raging whitecaps in a few hours. The sky may be very clear and the humidity extremely low, making it seem like the perfect day for an outing. A few near-shore whitecaps can warn the experienced sailor that a blow is imminent, or a forecast of moderate northeasterly winds below canyons is often a tip that stronger Santa Anas are brewing. Although warm Santa Ana winds can affect coastal waters, the cold Santa Anas forebode the worst weather.

Northeasterly Santa Ana winds often follow the northwest wind pattern. These winds result from a buildup of very high pressure over Nevada and Utah. They are accelerated by funneling through canyons and passes. Most of the time these winds blow only moderately strong, 30 to 50 km/hr, below the coastal canyons northwest of Santa Monica Bay and below Santa Ana Canyon in Orange County. However, on occasion they blow over all the coastal areas with speeds as high as 90 to 100 km/hr below some canyons. They can strike exposed offshore areas, such as Avalon Bay on Catalina Island, with sudden strong gales and 3 to 5-m high waves.

2. Nomenclature

There is much controversy over the correct term for the wind just described. Meteorological literature and the U.S. Weather Bureau use the term "Santa Ana" because the wind is most common to the pass and river valley of that name. The wind is also said to be named after General Santa Ana, the Mexican general whose cavalry stirred up clouds of dust during military campaigns. Most mariners seem to prefer the name "Santana" — corrupted from an Indian word "Santanta", which means "devil wind". Whatever its name, it does stir up clouds of dust during fall and winter months, brings havoc to harbors sheltered only from prevailing westerly winds, causes high fire danger in dry seasons, and creates locally severe turbulence to aircraft, being especially deceiving because of the cloudless sky and good visibility outside the dust areas.

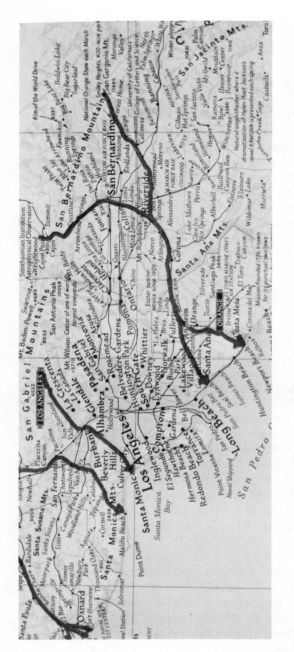

FIGURE 5.   Los Angeles-Hollywood area with the five inroads for Santa Ana winds: (1) Santa Ana Canyon; (2) San Gorgonio Pass; (3) Cajon Pass; (4) Placerita Canyon; (5) Santa Clara Valley.

Visibility in the Los Angeles area is better than average during this sequence of changes, since the normal summer temperature inversion deepens early in the process, while the arrival of cold air destroys the inversion entirely if the passing storm is sufficiently near and vigorous. Then Angelenos enjoy the clearest of air and can often see snowy mountain peaks 90 mi to the east.

3. Properties

It should be kept in mind that a dust storm such as the Santa Ana wind is different from a sandstorm. The sandstorm is a strong wind carrying sand through the air, the diameter of most of the particles ranging from 0.08 to 1 mm. In contrast to a dust storm, the sand particles are mostly confined to the lowest 3 m, and rarely rise more than 20 m above the ground. They proceed mainly in a series of leaps (*saltation*). Sandstorms are best developed in desert regions where there is loose sand, often in sand dunes, without much admixture of dust. They are due to strong winds caused or enhanced by surface heating and tend to form during the day and die out at night. In the U.S. weather observing practice, if visibility is reduced to between 5/8 and 5/16 mi, a "sandstorm" is reported; if visibility is reduced to less than 5/16 mi, the sandstorm is classified as "severe". This is rare with the Santa Ana which is mostly a dust storm.

The biometeorological effects are similar to those described for the foehn and sharav in this chapter and do not need special description. Ions and sferics are abundant and play havoc on the inhabitants of the Los Angeles-Santa Ana area.

4. Criminality

In discussing the Santa Ana winds, Miller[39] says that if people in Southern California are uncomfortable and irritable on days when there is a pronounced Santa Ana condition, it might be expected that an above-normal number of crimes would be committed during those days. A preliminary attempt to determine whether such a correlation existed in the Los Angeles area was therefore made by Miller.[39]

The dominant identifying characteristic of the Santa Ana is the low humidity, and it was the criterion used to identify the days during 1964 and 1965 on which the Los Angeles Civic Center experienced such a condition. The police department and the Sheriff's Office provided Miller with the number of homicides reported in their jurisdictions by day for 1964 and 1965, and since this is the ultimate crime of violence, it should, according to Miller, "constitute the most appropriate single index."

These two sets of data were combined to produce a single total number of homicides by day, and from this combined list the number of homicides reported for each of the 53 days during 1964 and 1965, when the civic center experienced a noon relative humidity of 15% or less, was obtained. In addition, the average number of homicides reported during both years by day of the week for the 8 months, September to April, during which Santa Anas typically occur were calculated; these figures (Monday 0.8, Tuesday 0.9, Wednesday 0.8, Thursday 0.7, Friday 1.0, Saturday 1.4, Sunday 1.2) being taken as the "normal". Each of the 53 actual daily number of homicides was then compared to determine the "departure from normal". This information together with the relevant weather data showed that, of the 53 days during 1964 and 1965 with a pronounced Santa Ana condition at Los Angeles Civic Center, 34 recorded an above-normal number of homicides, 3 had a normal number, and 16 had a below-normal number. The total number of homicides for all 53 days was 58, compared with a norm of 50.8. The departure from normal was therefore 7.2 or 14.2%. The period October 20, 1965, to October 26, 1965, had the longest sustained Santa Ana with a noon relative humidity of 15% experienced at Los Angeles Civic Center during the 2 years under consideration, and of these 7 days, 4 carried reports of homicides and 3 did not. However, Miller notes that "the total of 10 reported homicides was 47% above the 6.8 for a normal week."

Miller says, in conclusion, that it should be recognized that the time span for which data have been assembled is only 2 years, and there is no positive assurance that findings are not in part attributable to other currently unidentified factors. Accordingly, results of the study are believed to be indicative but not necessarily conclusive of the association between weather conditions (specifically the Santa Ana winds) and crime (specifically homicides) in the Los Angeles area.

5. Other Evil Winds

There are similar evil winds all over the world. The *California Norther* is a strong, very dry, dusty, northerly wind which blows in late spring, summer, and early fall in the valley of California or on the west coast when pressure is high over the mountains to the north. It lasts from 1 to 4 days. The dryness is due to adiabatic warming during the descent. In summer it is very hot.

The *Chocolatero* of the Mexican Gulf is not as hot as the Santa Ana wind, yet it is colored brown by sand particles.

The *Brickfielder* is a squally northerly or northwesterly wind in southern Australia. It comes in advance of the axis of a trough of low pressure advancing toward the east. Blowing from the dry interior of the continent, it carries much sand and dust, and in summer is very hot with day maxima exceeding 40°C. Sometimes the dust is thick enough to restrict visibility seriously. The Brickfielder is similar to some of the desert winds of Africa mentioned in this chapter.

The *Burster* or *Southerly Burster*, also called *Buster, Southerly Buster*, is a sudden shift of wind to the southeast in the south and southeast parts of Australia, especially frequent on the coast of New South Wales near Sydney in summer. It occurs in the rear of a trough of low pressure which is followed by the rapid advance of an anticyclone from West Australia. After some days of hot, dry, northerly wind (Brickfielder), heavy cumulus clouds approach from the south, the wind drops to calm and then sets in suddenly from the south, sometimes reaching gale force. Temperature at Sydney has fallen from 40 to 18°C in 30 min. The average summer frequency of Bursters at Sydney is 32. Similar winds are experienced in the east of South Africa, especially near Durban.

B. Foehn and Other Mountain Winds

1. Occurrence

Foehn (from the Latin word Favonius or from the Greek word Phoenix = warm south to southwest wind) is a common name in central Europe for a very warm, dry, and electrified wind which often blows in the mountain valleys of Switzerland and Tyrol. It starts with humid air from Italy rising against the high mountains of the Swiss Alps. The impact cools it so much that heavy precipitation results. The dried air which reaches the northern lower parts of the Alps warms up during this process and descends as a warm dry wind in the northern valleys. Similar winds are known in the western plains of the U.S. and Canada as *Chinook,* as *Autan* in central France or *Tramontana* in northern Italy, as *Zonda* in the Argentine, as *Koebang* and *Gending* in Java, and *Bohorok* in northern Sumatra. They are characterized by peculiar biological effects which also show up in the expression "losing one's tramontana" (being put out).

2. Evil Mountain Winds

Whenever a descending wind exists, its electrical impact on the population is feared and hated (Figure 6). This is also true for similar winds, such as the *Tauernwind* of

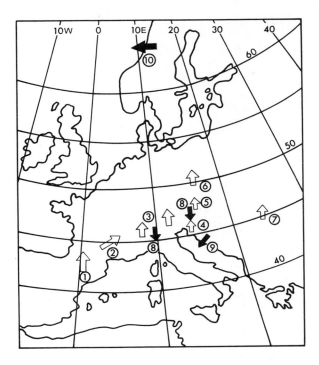

FIGURE 6.   Ill-tolerated fall winds in Europe. Transparent
arrows indicate warm winds, black arrows indicate cold winds.
(1) Vent d'Espagne (Pyrénées): warm; (2) Vent du Midi (Tarn
Valley): warm; (3) Foehn of the Alps: warm; (4) Junk of the
Karawanken: warm; (5) Pyrnwind of the Danube: warm; (6)
Riesengebirgswind of Silesia: warm; (7) Roteturmwind of Siebenbuergen: warm; (8) Mistral of the Rhône Valley, Maloya
wind, and Bise of the Alps: cold; (9) Bora of the Adria: cold;
(10) Bora of Norway: cold.

Salzburg, the *Pyrnwind* of the Danube, the *Jaukwind* of the Karawanken, the *Sued-
wind* of Innsbruck, the *Maledetto Levante* of Italy, the *Halmiak* or *Almwind* of the
Yugoslav Karst, the *Halne* of the Tatra, the *Roteturmwind* of Siebenbuergen, the *Au-
tun* of the Pyrénée, the *Vent d'Espagne* (France), and the *Vent du Midi* of Lyons. If
these winds are cold, they are less dreaded though rather unpleasant, e.g., the *Bise* of
the Lac Léman, the *Bora* (Latin: Boreas) of the Ligurian and Adriatic Coasts (see
Volume I, Chapter 2, Figure 19), the *Vardarac* of the Thessaloniki region, the *Mistral*
of the Rhone Valley and Provence, the *Maloya (Maloggia)* of the Engadin, the *Sansar*
(Sarsar, Shamsir, the "icy northwest wind of death") of Persia, the *Reshabar* of Kur-
distan, the *Norther* of Texas and Portugal, the *Norte* of north Mexico, Chile, and
Spain, the *Tehuanctepecer* of south Mexico, the *Papagayo* of Guatemala, the *Minuano*
of Brazil, and the *Pampero* of Argentina and Uruguay.

3. Three Foehn Variations
a. Southern Foehn

The foehn can be conceived as an offspring of the desert-born sirocco. The latter
arrives dry and hot in Sicily and Malta, yet by moving up to the north it becomes wet
and cold until it reaches the Alps. Passing the peaks, the foehn reaches down into a
valley and moves toward its opening. The cold air layer on the ground, which is less
than 100-m thick and is present in the cold season on the leeward lowlands, moves in

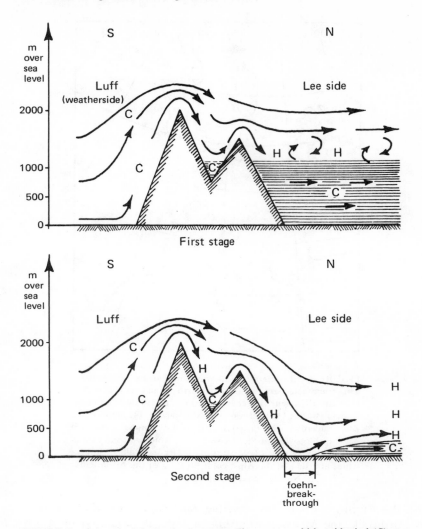

FIGURE 7.   Schema of foehn development. First stage: cold humid wind (C) ascending the Alps from the luff side, creating a warm rotor whirlwind on the lee side which ''brushes'' away the cold air layer (C) and produces sferics. Second stage: warm breakthrough on the lee side, heats up by adiabatic pressure, removes the cold air mass up to several hundred kilometers northward, and produces a surplus of positive ions.

the direction of the ''low'', i.e., toward the north, forming a passive north wind. This wind creates the returning branch of the turbulent current formed during the prefoehn stage as a whirl-rotor at the site of the foehn inversion between the cold ground air and the foehn current above, as shown in Figure 7. In transverse valleys, the foehn goes down to the ground, where a north to south course of a longitudinal valley leads to the transverse valley, for example, from the Brenner Pass north toward Innsbruck (Figure 8). The longer the low pressure area (cyclone) is present, the longer the foehn will last. Concomitantly, there is a well-developed ''high'' on the east side, especially in the cold season. The warm air current brings maritime air from southwest Europe across the Alps far to the north. Thus, the foehn is the Alpine modification of an approaching south current on the frontal side of a cyclone.

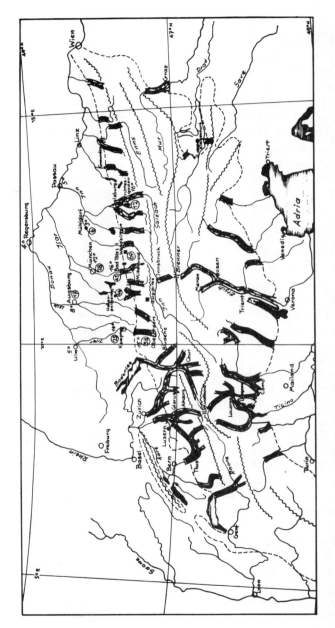

FIGURE 8.   Foehn valleys in the Alpine region. The southern wind moves through them mainly in a northerly direction and then descends as a hot adiabatic wind, increasing the incidence of complaints mentioned in this chapter by the presence of sferics and ions.

b. Free Foehn

In addition to the common *Southern Foehn* there also exists a *High-Altitude Foehn*. The anticyclonic foehn, also called *Free Foehn* or *High-Altitude Foehn,* is a completely different meteorologic phenomenon with descending air currents, but since these are hard to differentiate for the layman, it has become common to call both phenomena foehn.

c. Northern Foehn

When the winds come in from the north of the Alps, they may descend in the Italian area of the Alps as *Northern Foehn.* Most reports indicate that there are fewer biologic influences of the north foehn than of the south foehn. The north foehn is usually described as cooling, refreshing, and vitalizing, and this is consistent with the change of the weather to cool-dry. It is possible that the north foehn in the colder season acts like the foehn in the lower layers, but this depends upon the intensity of meteorologic developments, including the temperature of the air masses on arrival. In most cases, however, the north foehn appears as a cold wind (Bora). The period just preceding the arrival of the north foehn is characterized by high ion and sferic counts, considerably increasing rheumatic and scar pain as well as neurovegetative disturbances, which decrease gradually after its arrival, yet there is a striking increase in spastic conditions. All these complaints are typical of serotonin release.

4. Foehn Properties

Wherever air is compressed, heat is created. This is what happens with the foehn which is due to a local depression (cyclone) that does not receive any energy from the vicinity. Therefore this cyclone has been called an *adiabatic* one, i.e., an air mass with no heat permeation from the surrounding atmosphere. Thus, two well-separated air masses exist in the foehn area and their boundaries create an unpleasant zone of friction with high electricity charges. This zone is in permanent motion creating the so-called *Kelvin-Helmholtz Waves.*

Foehn is characterized by the following properties:

1. Electrical charge of sferics (1 to 100 kHz impulses of 1/1000 sec) and ions (mostly positive 4000/cm³), the occurrence of the latter being disputed in Germany because they hardly arrive at the Bavarian side.
2. Adiabatic heat increase of 1°C/100 m.
3. Loss of relative humidity from 100% down to 10 to 20%.
4. Turbulence due to the Kelvin-Helmholtz Waves.
5. Formation of typical clouds such as the Foehn Wall on the peak of the mountain (*stratocumulus*) which releases smaller lens-shaped clouds (*altocumulus lenticularis*) through a narrow opening, the *Foehn Window* (see Volume I, Chapter 2, Figure 35).
6. Special fall winds on the lee side (Figure 7).
7. Completely blocked view of the luff side of the mountain.
8. Unusual visibility on the lee side of the foehn.
9. Barometric pressure variations of 0.1 to 1 mbar which can be measured by the Variograph®.[40] This instrument shows that at a frequency period of 4 min there are rarely any pressure changes; however, in foehn conditions the amplitudes beyond the 4-min range are increased, whereas the amplitudes below the 4-min range are decreased.
10. Pressure variations are due to the Kelvin-Helmholtz Waves; thus these waves determine three factors which can provoke the foehn complaints: sferics, ions, and pressure oscillations.

11. There is general agreement that some diseases are exaggerated by foehn: ulcer perforations and hemorrhages, thromboembolism, migraine, depression conducive to suicide.
12. The effect of foehn on other diseases is doubtful: edema, hypothyroidism, hypoadrenalism, apoplexy, hypertension and hypotension, myocardial infarction, appendicitis, gallbladder, and renal colics.

5. Foehn Disease

The manifold complaints connected with appearance of the foehn are not due to dry heat alone. It is reasonable to assume that they affect people who are sensitive to one of the three electric foehn attributes which may well arrive 1 to 2 days before the foehn wind because it takes them only seconds to spread, whereas the foehn wind moves slowly. These three attributes are (1) sferics; (2) ions; and (3) biometric pressure variations.

Many researchers have studied these aspects and their findings are so contradictory that no acceptable explanation has been forthcoming. For this reason is seems probable that the 30% of "Foehn Prophets" in a given population are sensitive to one of these attributes only. This would explain why researchers of one attribute arrived at different conclusions. The solution is bound to come through the construction of three different apparatuses which would neutralize the above-mentioned three foehn attributes, a *pium desiderium* not yet realized.

Foehn disease itself can be conceived as due to three main disturbances and three contributing factors.

The main disturbances are due to:

1. Serotonin release by electrical charges occurring 1 to 2 days before arrival of the foehn
2. Adrenal exhaustion due to the constant stress imposed on the adrenal gland, getting worse every year
3. Thyroid hyperactivity provoked by the hot dry wind and its pressure variations

The three contributing factors are

1. Vegetative dystonia which makes the body over-sensitive to every stress
2. Hypertension with its high epinephrine level affecting the sympathotonic patient
3. Obesity which is conducive to heat accumulation and reduced heat dissipation

The special ill-effects of the foehn are similar to the stress reactions described earlier in this chapter. They rest mainly on the release of serotonin, histamine, or thyroxine. There are however additional disturbances in blood capillary constriction resulting from the lack of the catecholamines (epinephrine and norepinephrine). This has the following consequences:

1. It may facilitate the penetration of microorganisms or viruses and the toxic substances created by them and of other toxic substances through the skin or mucus membranes of the nose and throat.
2. Oxygen can pass rapidly out of the blood plasma through the capillary wall and tissue into the tissue cells. With increased permeability of the membranes the process is accelerated. In other words, it favors oxygen intake and CO_2 release. This results in a hyperventilation syndrome which binds calcium and promotes cramps or tetany.
3. Increased permeability of the kidney filtration units (*Malpighi's* bodies and *Bowman's* capsules) allows escape of albumen and blood cells into the urine.

Foehn research is a riddle which has until now defied the efforts of thousands of physicians and meteorologists. It has, however, contributed to the development of a fruitful conception: the weather phases described in the next section.

6. European Weather Phases

The foehn and its impact on human well-being has resulted in the conception of six weather phases, the sequence of which is rather stable. This was the work of H. Ungeheuer[41] and H. Brezowsky[42] who headed the Medico-Meteorological Research Station at Bad Toelz in Bavaria. Their classification takes special account of the weather preceding and following a foehn as follows (Table 7):

Phase 1: cloudy fair weather — Average nice weather due to a steady "high" with partly cloudy sky and usual daily variations in temperature and humidity. Barometric pressure and sferics counts are rather high. This weather has a cool, dry, refreshing effect on the human body and shortens reflex reactions due to epinephrine release.

Phase 2: sunny fair weather — Very nice weather due to shift of the "high" in an easterly direction results in decreasing cloudiness, slight fall in barometric pressure and rhythmic changes of the temperature-humidity curves with increased air ion content. This weather is experienced as refreshing and pleasantly warm, but arterial spasms may produce migraine, apoplexy, nervous reactions, and suicide traits, due to the hyperventilation syndrome with its concomitant hypocalcemia.

Phase 3: foehn fair weather — Extremely fine weather due to an incoming weather front from the west; it is accompanied by lenticular and cirrus clouds, falling of barometric pressure and rising of temperature, ion, and sferics counts caused by influx of warm dry air. The weather is usually experienced as unpleasantly warmish or warm and provokes sleeplessness, tension, irritation, hypertonic reactions of heart and circulation, apoplexy, all kinds of colics, migraine, low concentration with increased accident rate, and suicide tendency. This phase is due to serotonin over-production and may precipitate thromboembolism.

Phase 4: changing weather —Initial weather change due to passing of cold fronts produces rapid appearance of rain and thunderstorm clouds, steep fall in barometric pressure, temperature remaining high but humidity increasing. The weather is experienced as sultry and oppressive with high sferic counts. Therefore, it provokes depression, spastic reactions of the intestines and blood vessels, culminating in apoplexy, epileptic seizures, and shortened reflex time. Moreover, thromboembolism, heart infarcts, ulcerative bleedings, polyarthritic pains, glaucoma attacks, lumbago and sciatica attacks ("Hexenschuss"), and premature delivery occur.

Phase 5: rainy weather —Complete weather change due to influx of cold air under a heavily clouded sky with rain showers; after the passage of the front barometric pressure rises, temperature falls steeply and humidity is rather high. Electric fields show a high positive charge. Therefore, the weather is experienced as uncomfortable, penetratingly cold, and provokes spastic reactions of the arteries and the intestines, angina pectoris, and spastic bronchitis.

Phase 6: balanced weather — Initial weather improvement due to return to steady weather conditions. The sky becomes less cloudy, blue patches appear, humidity falls; temperature is still low but during the day the temperature rises. Electrical field charges oscillate, therefore the weather is experienced as pleasantly stimulating but not warm. Diseases are rare, yet rheumatic pain may occur.

This six-phase weather cycle (Table 7) has now been adopted in West Germany, East Germany, and Switzerland, and a similar classification could be worked out for many other areas of the temperate zone where mountains create hot, dry, electrified fall winds. The German Weather Service (Frankfurt-Offenbach) has issued a special booklet on the medical implications of these weather constellations. However, a cursory

Table 7
SYNOPSIS OF THE SIX ALPINE WEATHER PHASES

Phase	1	2	3	4	5	6
Weather	Fair weather	Sunny weather	Foehn weather	Weather front	Rainy weather	Balanced weather
Clouds	Cumulus or cirrocumulus	None or cirrus	Altocumulus lenticularis	Cumulonimbus or cirrostratus	Nimbostratus or stratocumulus	Altocumulus or altostratus
Relative humidity	Medium	Medium	Low	Medium	High	Medium
Temperature	Medium	Warm	Hot	Cold	Medium	Cold
Psychic effect	Refreshing	Pleasant	Depressive	Oppressive	Uncomfortable	Stimulating
Diseases prevailing	—	Migraine Apoplexy Epilepsy	Sleeplessness Irritability Tension	Thromboembolism Apoplexy Epilepsy Polyarthritis Lumbago Abortion	Angina pectoris Spastic bronchitis Spastic colitis	Rheumatic pain Phantom pain
Air electricity	Normal electrofields	Normal electrofields	Sferics Ions	Sferics Ions	Sferics Ions	Electrofields changing

attempt to establish in nine "weather-sensitive" patients a correlation between the aforementioned six phases and urinary 5-HT and 5-HIAA did not yield practical results because the weather-sensitivity of the patients was ill-defined.[43]

C. Sharav and Other Desert Winds
1. Definition
Sharav, Khamsin, Sharkiye = Sirocco (Italian) denotes a condition of the weather which is much talked about in the Mediterranean countries. The term *sharav*, used in Israel, has been taken from the desert weather already referred to in the Bible (Isaiah — 49:10 evil wind, 25:7 fata morgana).

Meteorologically this weather situation arises at certain times of the year, usually in spring and fall under three different meteorological conditions (Figure 9).

Cyclone — When a succession of "lows" moves eastwards along the North African Mediterranean coast. When they reach the Sudan, southeasterly to easterly winds (anticyclones) blow up, which drive the hot desert air — often with a dust cloud ceiling — from the Sahara desert into the southeastern part of the Mediterranean basin. This sharav lasts 1 to 2 days.

Trough — When an elongated area of "lows" extending from the Red Sea to the Mediterranean carries in hot, dry winds from the deserts of Arabia, Iraq, Syria, Jordan, and Iran. They do not usually carry much dust or sand. This sharav lasts up to 1 week.

Anticyclone — When a "high pressure ridge" develops over the Middle East. This occurs mostly in summer or winter, being associated with cloudless, hot, and extremely dry weather. The air coming in from above does not win or lose its quality; it is therefore called "adiabatic" (impermeable) and is warmed only by pressure from above like the foehn. It may last for several weeks and is mostly accompanied by a complete absence of winds (occlusion). This form of the summer-sharav is hard to bear as it increases summer heat and lasts up to 3 weeks. The winter-sharav occurs less often and may not appear at all (winter 1979—1980). It is a pleasant diversification to the winter cold and rain, and only weather-sensitive people suffer from it.

2. Meteorology
Meteorologically, the sharav has nothing to do with the foehn known in Europe. What they do have in common though is that, seen from the psychological point of view, they are held responsible for everything from migraine to heart attacks, from bad moods to car accidents — and rightly so.

In Arabic, Sharkiye = Sirocco means "easterly", and Hamsin (Khamsin) means "fifty", a euphemism that buoys up the sufferer with optimism, that when he has served his 50 Hamsin days it will all be over. There are, however, years in which up to 50% of the days are recorded as Hamsin days. In Israel this sharav wind is noticed mainly in the mountainous areas (Jerusalem, Safed). There it creates a heat spell, accompanied by stronger or weaker easterly winds and characterized by the exceptional dryness of the air (down to 0% relative humidity). In the coastal plain further to the west, where normally a fresh western sea breeze prevails, the damp air then comes to a standstill *(occlusion)* and the sea with it. This situation may last from only 1 day to, though rarely, 3 weeks. It returns to normal when a cold westerly wind comes in, either abruptly within a quarter of an hour (similar to a thunderstorm) or gradually over several days.

3. Biometeorology
The sirocco and sharav differ from the foehn. They are, however, similar both in their physical properties and in their pathophysiological effect. They are dry, warm

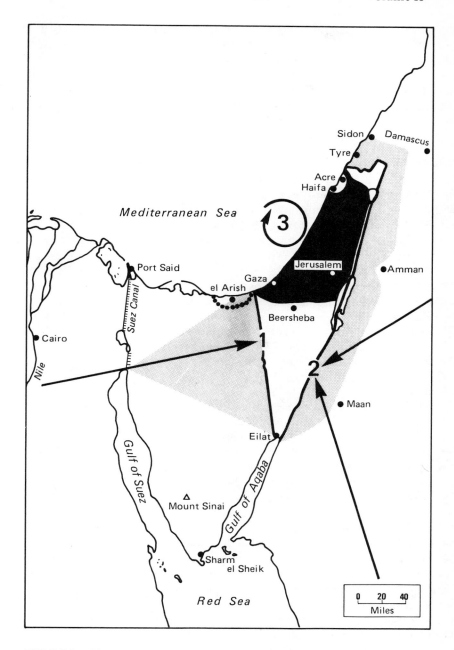

FIGURE 9. Three variants of sharav (khamsin): (1) cyclone from Sahara in spring or autumn, lasting 1 to 2 days: (2) trough from Jordan, Arabia, Iraq, Syria, and Iran in spring or autumn, lasting 1 day to 1 week; (3) anticyclone from above (adiabatic) in summer or winter, lasting 1 to 3 weeks. Densely populated area is marked in black.

winds with a strong positive and very low negative ion concentration and varying sferics count. They are best characterized by the following definition: dryness — below 30% relative humidity (down to zero); heat — more than 10 to 15°C above the average for the season (up to 45°C); electricity — 1000 positive and negative ions per cubic centimeter of air (positive ions up to 5000 and negative ions up to 4000/cm³ sometimes much less or none at all); sferics — high at the beginning and at the end, but sometimes absent during the heatwave. Yet, each one of the three sharav variants has its own pattern of electrical phenomena; e.g., the trough-sharav does not show sferics, because it loses its sand particles when it passes the Jordanian mountains.

It is not the dry heat which affects man's well-being with the psychic symptoms, but the electricity in the air, since dry heat is always accepted as pleasant. However, both foehn and sharav release serotonin and irritability spreads in central Europe as in the Holy Land, aggressiveness and listlessness increase, and, wherever they blow, accident and suicide figures shoot up.

4. Other Desert Winds

Similar observations are also reported for other evil winds indicated in Figure 10: *Leste* or *Leveche* (Morocco); *Harmatan* (West Africa); *Chergui, Chom, Arifi, Chili, Chichile* (Algeria); *Chili* (Tunisia); *Ghibli* (Tripolitania and Libya); *Khamsin, Haboob, Aziab* (Egypt); *Shamal, Sharqi, Reshabar* (Iraq); *Seistan* (Iran); *Simum* (Syria, Jordan); *Simoon* or *Sharkiye* (North Africa); *Sirocco* (Italy); the *Meseta-Winds* viz. *Levante, Gerona,* and *Tramontana* (Spain); *Lips, Meltemia* = Homer's *Etesiae* (Greece); *Klokk* (Malta); *Thar Winds* (Rajasthan, New Delhi, Agra, India); *Northern Winds* (Melbourne); *Norther* (Southeast Australia); *Canterbury North-Western* (New Zealand); *Zonda* (Argentina); *Santa Ana* (California); *Arizona Winds* (Arizona).

There are some excellent accounts of the sharav by Selma Lagerloef in *Jerusalem,* of the Khamsin by Albert Camus in *The Stranger,* and of the sirocco by W. Brydone in *Journey to Sicilia and Malta in 1770.* Thomas Mann has given a touching description of the Tramontana in *Der Tod von Venedig.*

5. Alarm Reaction

Basically the warm and dry electrified atmosphere exerts a strain on the body, to which it responds, in accordance with Selye's theory of stress, with an alarm reaction. The adrenal medulla produces increased epinephrine (adrenaline) whereby the vessels, which had become dilated as an initial reaction to the heat, again become restricted and perspiration is reduced (Figure 11). This normal reaction is usually sufficient in the case of residents to compensate for the dry and warm air stress. Sometimes, these winds even have a euphoric effect if epinephrine (adrenaline) production is adequate. This is how the American Bishop Pike sadly lost his life in 1969, when he ventured one sharav day into the Judean Desert. He wanted to follow in the footsteps of Jesus and John the Baptist, but he did not have their stamina.

Over the years, however, the ability to adapt to dry heat and sunshine vanishes because the organism cannot always initiate the alarm reaction and can no longer hold out during extended hot dry periods. The propensity of the adrenal glands and hypophysis to secrete is gradually reduced, as we can often demonstrate in the urine of native and long-settled Israelis (Table 2). In addition to this, protracted exposure to sun radiation can evoke the Janus-faced skin effects shown in Volume I, Chapter 2, Figure 7.

6. Adrenal Exhaustion

Originally, with the aim of discovering suitable antifoehn preparations, we arranged for ten scientists from our department to have their urine regularly tested for hormones and electrolytes. At the same time, 800 people who themselves suffered particularly from the desert wind, similarly provided regular samples. The findings from sharav periods were compared with those of normal weather conditions as well as with those of control persons in areas free of sharav.

After 4 years the results of the study, which was sponsored by a grant from the U.S. National Institutes of Health, led us to the conclusion that there would hardly ever be

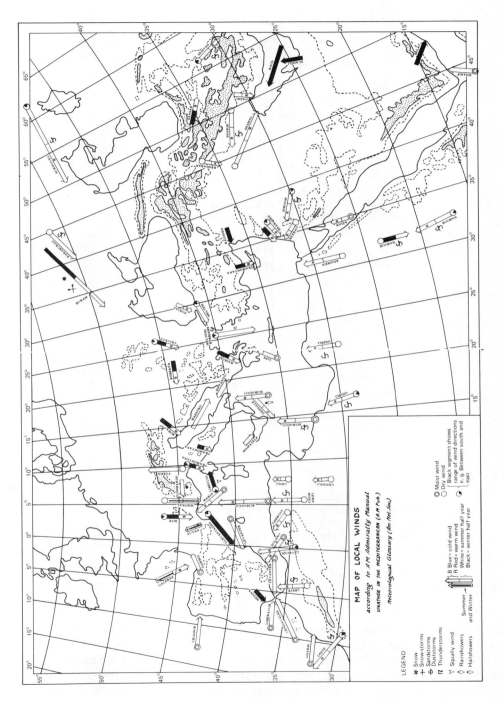

FIGURE 10.   Map of local winds in Eurasia and Africa.

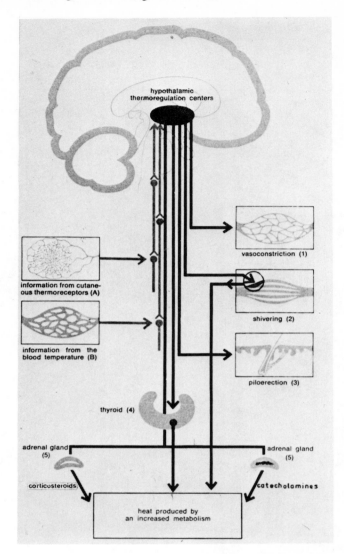

FIGURE 11.   Diagram illustrating the mechanisms of thermoregu-
lation in the human body. The hypothalamus contains the thermo-
regulation centers which are connected by hormonal releasing factors
to the hypophysis. The latter governs by its "tropin" hormones the
main two endocrine glands involved: thyroid (4) and adrenal gland
(5). Information is added from the cutaneous thermoreceptors (A)
and from the blood temperature sensors (B). The heat buffering
mechanisms are provided by cutaneous vasoconstriction (1), muscle
shivering (2), and skin piloerection (3).

a universal therapy of weather sensitivity. The urine findings pointed to no fewer than
three clearly differentiated reaction mechanisms, which, separately or simultaneously,
and in ever-varying strengths, can come into play. They are described in Table 2. Long-
time residents reacted quite differently to sharav stress than did newcomers or tourists.
It confirmed the old experience that foehn or sharav sensitivity does not become evident until years after one has moved into the area.

7. Sodium Loss, Potassium Excess, and Dehydration

Two further processes come into play: in the dry warm air, fluid loss through per-

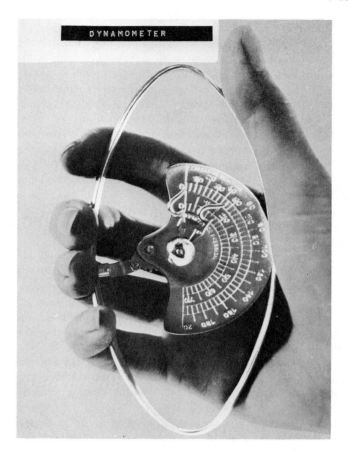

FIGURE 12. The Dynamometer Test. The instrument indicates the grip strength of the hands, measuring the force of muscular contraction, due to sodium, potassium, and acetylcholine reserves. This simple instrument allows the differential diagnosis of weakness and fatigue due to heat stress: the latter — being due to lack of catecholamines — does not reduce the grip strength, but excess potassium or over-exposure to insecticide pollution reduces it considerably, due to lack of acetylcholine and minerals at the neuro-muscular junctions.

spiration can rise from the normal 25% of the overall fluid flow to 50%. With the perspiration, more sodium is lost (up to 20 g/day) and the cells seek to compensate for the electrolyte deficiency by circulating more potassium. Since, in larger quantities, potassium can have a toxic effect on the muscles of the heart, the flood of potassium in the blood can explain not only the general foehn and sharav weakness experienced by healthy people but, above all, the recognized decompensation of those with heart ailments during foehn and sharav conditions.

At the same time the loss of sodium causes increased secretion of glucocorticosteroids in order to eliminate the potassium from the blood through a raised sugar level, while at the same time fewer androcorticosteroids are produced. The androcorticosteroids (17-KS) are the true stress corticosteroids and their deficiency deprives us of the ability to cope with heat stress. Their decrease is probably to be explained by a deficiency of the adrenal cortex overburdened by increased production of glucocorticosteroids. The typical symptoms of this threefold adrenal dysfunction together with the loss of sodium and the compensatory hyperpotassemia are bradycardia, muscle weakness, and impaired ability to concentrate (Figure 12).

8. Sferics and Ionization Release Serotonin

With the sharav as with the foehn, sferics and positive ionization of the air seem to play a special role. Ionization of the air with a sudden change in dry weather is the correlate of lightning in the case of a sudden change in wet weather. Lightning produces sferics (electromagnetic impulses) and ions (electrical charge of air, oxygen, and water). Both arise through friction between different layers of clouds or air-dust particles. The ions in the air are also produced at great heights by cosmic radiation and at lower level by emanation of radon, thoron, and actinon from the surface of the Earth. Recent research has shown us that the ion activity of the activated electrically-charged oxygen in the inhaled air has a considerable effect upon lung functions: oxygen in the air with strong positive ionization is taken up along with normal oxygen via the lung alveoles and avidly binds to the hemoglobin of the red blood corpuscles; moreover it lowers the partial oxygen pressure in the alveoli, while the partial carbon dioxide pressure goes up. The reduced respiratory capacity in turn causes mental ability and resistance to stress to drop significantly, and over and above that it produces another stress hormone — serotonin. This is released from the blood platelets (thrombocytes) when they meet positive oxygen or water ions in the lung alveoles.

Sferics and positive ionization always precede the weather front by 1 to 2 days because electricity moves faster than air. This accounts for the particular presensitivity to weather of many people. These people maintain, and rightly so, that they are weather prophets.

Preponderantly positive ionization of the oxygen molecules is, of course, not only found in foehn-type weather. It is also a feature of foggy and polluted city air, because the water drops and the dust particles are mainly positively charged and thereby quickly neutralize the negative oxygen ions. In cities, after extended periods of calm weather, there prevails — to a certain degree — a "Static Foehn" with low ion count. Negative ionization applied to the air conditioning of many modern high-rise blocks, in which the positively charged air to be breathed is enriched with negative ions, thus actually represents a social step forward.

In the same way, patients who are sensitive to sharav can today be treated prophylactically with negative ions from a single ionizer which neutralizes the positive ions so as to a large extent to avoid these troubles.

III. THERAPY OF CLIMATE AND WEATHER SENSITIVITY

A. Prophylaxis

General remedies are available against Santa Ana, foehn, and sharav winds arising from the characteristics of the various physiological reactions to the pathogenic weather (Figure 13): abundant drink in order to correct the fluid loss, sugar against hyperpotassemia, and salt to compensate for the loss of sodium through perspiration. Salt alone causes thirst, but salted herring with its many amines is a more convenient form of salt medication.

Regular comparative urine tests show three groups of reaction in patients sensitive to the weather (Table 1).

The biometeorologic changes due to changes of weather and climate in body glands, fluids, and neurohormones are so diverse that symptomatic therapy as devised until now is not very helpful. The weather-sensitive patient with his three different reactions to climate changes (serotonin release), heat (thyroxine stimulation), and dryness (catecholamine deficiency) is described earlier in this chapter. This intricate condition requires prevention, relief, and eventually cure. The specific methods developed by us are presented in the following paragraphs. Their individual adaptation to the needs of the weather-sensitive patient controlled by neurohormone urinalysis requires much ob-

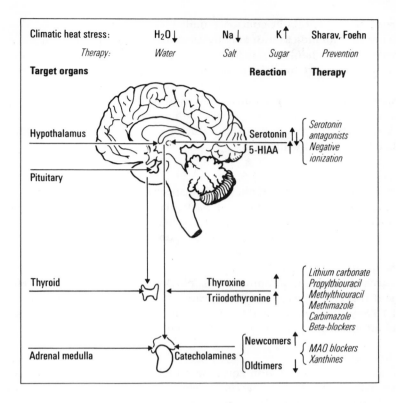

FIGURE 13. Scheme of treatment for patients suffering from sharav or foehn stress. Replacing water loss by cool drinks, sodium loss (Na) due to sweating by salt intake (salt herrings), and reducing potassium excess (K) by sugar intake which diminishes blood potassium. The three target organs of heat stress are the hypothalamus with the pituitary, the thyroid, and the adrenal gland, especially its medulla. Every heat stress reaction requires its specific therapy, as shown by arrows and list of drugs.

servation, patience, and cooperation, yet results are highly rewarding since they restore full work activity to an incapacitated patient.

B. Management

Comparative urinalysis of neurohormones, minerals, and the metabolic hormones enables us to establish the division into three main reaction groups of the manifold weather complaints. These tests compare the daily secretion on normal days with that on Santa Ana, foehn, or sharav days and allow an adequate indication for treatment (Table 2).

The following 12 parameters must be investigated in the urine (Table 1):

1. Serotonin and its metabolite
2. Hydroxyindole acetic acid (5-HIAA)
3. Epinephrine (adrenaline) (adrenal catecholamine)
4. Norepinephrine (noradrenaline) (adrenal catecholamine)
5. 17-KS (adrenal androgens and stress hormones)
6. 17-OH (adrenal glucocorticoid metabolites)
7. Sodium and its replacement by
8. Potassium, which when high comes from the body cells

9. Creatinine as a metabolic control of daily urine measurements
10. Histamine as an indicator of allergic reactions or hyperthyroidism
11. Thyroxine as a reliable parameter of thyroid function
12. Diuresis to measure the daily urine output

Carrying out clinical examinations in addition to these urine tests makes it possible to diagnose to which of the three typical weather-sensitive groups the patient belongs. In accordance with the results, a specific therapy can then be initiated whose success is controlled by means of urinalysis.

C. Treatment

The treatment of weather complaints with special drugs gives excellent results. Specific medicines which could become of particular importance for foehn therapy are basically different, corresponding to the three main reaction groups. The effect of all treatments is regularly controlled by neurohormone urinalysis.

1. Catecholamine Deficiency

This is treated with monoaminoxydase blockers (MAO inhibitors). However, all medicines of this type must be used with care. (A quarter of a tablet in the morning is sufficient.) They allow the adrenal to recover and the patient to get new strength. Their names and mechanism of action are given in Figure 14.

2. Serotonin Release

This is prevented by ionization equipment and treated with serotonin antagonists. Experience with the serotonin antagonists which we are using at the present time (Sandomigran, Periactin, Deseril, Dihydergot, Danitracene, GC-94), shows that preference varies with each patient. Their mechanism of action and structure are given in Figure 15. It is recommended with all serotonin antagonists to start with small doses at nighttime because of their sedative effect. Patients need not take more than one fourth to one half tablet once a day. In the course of the treatment, the sedative effect of the serotonin antagonists disappears and the specific antiserotonin effect prevails.

Postitive air ions and sferics release serotonin. Its irritative effects have been described in Volume I, Chapter 4. Drug treatment should preferably be prophylactic by using antiserotonin drugs. A list of these is given in Figure 15. A perusal of the different structures will reveal that all of them are built in a way which simulates the structure of serotonin, ring — C beta — C alpha — amine (Figure 16). This allows the ten compounds to compete with serotonin for its receptor. Therefore they should be given prophylactically before serotinin has been released and arrived at its receptor. Moreover, some of them may produce side effects which should be avoided by careful dosing, preferably only quarter tablets.

Ergotaminetartrate —Available as Ergomar® sublingual tablets (Fisons), Ergostat® sublingual tablets (Parke-Davis), Gynergen® ampoules and tablets (Sandoz), and Ergotamine Medihaler (Riker). Protracted use may produce arterial spasms of brain and limbs.

Dihydergotamine mesylate —Available as Dihydergot (Sandoz) and Ergogine (Abic). Protracted use has no side effects.

Methysergide —Available as Sansert® (Sandoz US) or Deseril (Sandoz Europe). It produces unpleasant feelings at the beginning and tolerance during protracted treatment which necessitates increase of dosage and eventually induces retroperitoneal fibrosis with pain in back and legs.

Ergoline — Available as Sermion (Farmitalia). It is not very active.

Metergoline — Available as Lyserdol (Farmitalia). It is not very active either.

Monoamine receptor and
Monoamine oxidase at ④

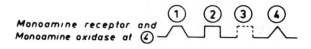

MONO-AMINE HORMONES:

1   NOR-ADRENALINE
$HO-\langle\bigcirc\rangle-\overset{}{C}H-CH_2-NH_2$
HO OH

2. ADRENALINE
$HO-\langle\bigcirc\rangle-\overset{}{C}H-CH_2-NHCH_3$
HO OH

3   SEROTONIN
$HO-\langle\bigcirc\hspace{-4pt}\bigcirc\rangle-CH_2-CH_2-NH_2$

M.A.O. BLOCKERS:

4. IPRONIAZID
(Marsilid, Euphozid)
$\langle N\bigcirc\rangle-\overset{}{\underset{O}{C}}-NH-N\overset{H}{\underset{\overset{|}{C}-H}{\diagdown}}\begin{smallmatrix}CH_3\\CH_3\end{smallmatrix}$

5. PHENELZINE
(Nardil, Phenodyn)
$\langle\bigcirc\rangle-CH_2-CH_2-N\overset{H}{\underset{NH_2}{\diagdown}}$

6. PHENIPRAZINE
(Catron, Cavodil)
$\langle\bigcirc\rangle-CH_2-\underset{CH_3}{\overset{|}{C}H}-N\overset{H}{\underset{NH_2}{\diagdown}}$

7. TRANYLCYPROMINE
(Parnate)
$\langle\bigcirc\rangle-CH-\underset{CH_2}{\overset{}{C}H}-NH_2$

8. NIALAMIDE
(Niamide)
$\langle N\bigcirc\rangle-CO-NH-NH-CH_2$
$\langle\bigcirc\rangle-CH_2-NH-CO-\overset{|}{C}H_2$

9. ISOCARBOXAZID
(Marplan)
$\langle\bigcirc\rangle-CH_2-NH-\underset{}{\overset{}{N}H}$
$CH_3\underset{O}{\overset{N}{\boxempty}}-C=O$

10. MEBANAZINE
(Actomol)
$\langle\bigcirc\rangle-\underset{CH_3}{\overset{|}{C}H}-NH-NH_2$

11. PARGYLINE
(Eutonyl)
$\langle\bigcirc\rangle-CH_2-\underset{CH_3}{\overset{|}{N}}-CH_2-C\equiv CH$

12. IPROCLOZIDE
(Sursum)
$Cl-\langle\bigcirc\rangle-OCH_2-CO-NH$
$(CH_3)_2-CH-\overset{|}{N}H$

FIGURE 14.   Treatment of adrenal medullary deficiency resulting from adrenal exhaustion after prolonged exposure to heat. Skin, nerve, and arterial tissue have a receptor for the adrenal monoamines — noradrenaline (1) and adrenaline (2) — which provide us with stamina, a feeling of well-being and the ability to cope with heat stress. They fit into this receptor just as does serotonin (3), the irritating hormone of the brain and intestines carried by the blood platelets (thrombocytes). The MAO blockers also fit into the same receptor and at its fourth "triangle site" inhibit the monoamine oxidase which normally terminates adrenaline action by oxidative destruction. A minidose (one fourth tablet) of a MAO blocker preserves the reserves of noradrenaline and adrenaline (the so-called catecholamines) in the patient and cures fatigue, adynamia, apathy, exhaustion, lack of concentration, confusion, and even depression, without affecting serotonin release at the low dose. In the meantime the patient's exhausted adrenal gland can recover. The MAO blockers, when given in high doses (3 to 6 tablets daily), render the patient sensitive to tyramine, a wakeamine contained in yellow cheese. Therefore the sharav- or foehn-patient should take only minidoses which are completely innocuous. The medication of choice is Marplan (9).

SEROTONIN ANTAGONISTS

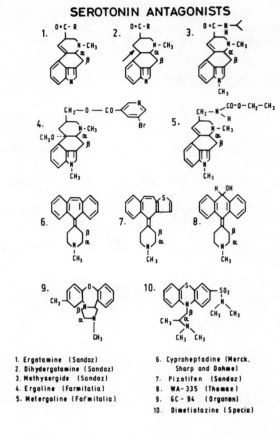

1. Ergotamine (Sandoz)
2. Dihydergotamine (Sandoz)
3. Methysergide (Sandoz)
4. Ergoline (Farmitalia)
5. Metergoline (Farmitalia)

6. Cyproheptadine (Merck, Sharp and Dohme)
7. Pizotifen (Sandoz)
8. WA-335 (Thomae)
9. GC-94 (Organon)
10. Dimetiotazine (Specia)

FIGURE 15.   List of ten serotonin antagonists; their effect is both antiserotoninic and antihistaminic, therefore they are also called antamines. They may use a common receptor, including that for norepinephrine (noradrenaline) as shown in Figure 16.

Cyproheptadine — Available as Periactin® (Merck, Sharp & Dohme), Oractin (Assia — Israel), Nuran (Sharp & Dohme). It is extremely active but induces dizziness and increased appetite. It should therefore be given in quarter doses before retiring.

Pizotifen — Available as Sandomigran (Sandoz). It is eight times more active than cyproheptadine and it, too, elicits dizziness and increased appetite. Use in quarter doses at nighttime is advisable.

WA-335 — Not yet being marketed. It is produced under the name of Danitracene by Thomae in Biberach (Germany). It is a very useful drug, active in doses of 0.5 mg, yet it may have the typical antiserotinin side effects, such as dizziness and increased appetite; moreover it may induce euphoria. All this can be overcome by low dosage.

GC-94 — Not yet being marketed. It is produced by Organon® (Holland). It is active in doses of 5 mg and has the advantage of being relatively free of the sedative side effect and the tendency to weight increase

Dimetiotazine — Available as Migristene (Specia). It has an unpleasant sedative side effect as do all phenothiazine derivatives. Moreover its activity is reduced by the presence of a methyl group on the C alpha, in the side chain.

3. Intermittent Hyperthyreose

Even in its slightest form, this complaint must be treated specifically: administration

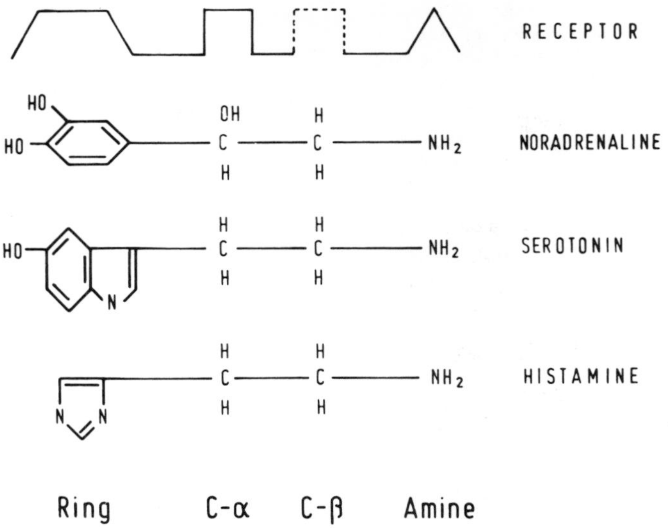

FIGURE 16.   Biogenic amines, norepinephrine (noradrenaline), serotonin, and histamine, and their partly identical cell receptor which can be blocked by antamine drugs (Figure 15).

of thiouracil twice a day can cure the patient of his over-sensitivity to sharav, though the treatment may require some weeks (with leukocyte control). We have therefore introduced treatment with 1 to 3 × 250 mg/day lithium carbonate for thyroid suppression. It works quickly and does not need blood control if used for 1 to 2 weeks only. Small doses of beta-adrenergic blockers should be added to slow down the pulse (Figure 17).

4. Treatment by Enzyme Induction

Recently we introduced a novel method to overcome the upset hormonal release in weather-sensitive patients.[44] The method uses proxibarbal (Axeen-Hommel). It is a barbiturate, devoid of any sedative effect but imbued with the typical barbiturate propensity to induce enzymes (Figure 18). A treatment lasting 1 to 2 months (3 × 100 mg daily) increases monoamine oxidase which destroys any serotonin surplus and rids the weather-sensitive patient of his serotonin complaints, especially migraine.

Treatment of intractable allergic diseases by enzyme induction represents a new method which should be more widely used. It has proved itself for many years in the treatment of unspecified cases of migraine. The first phase of this investigation, measuring the effect of proxibarbal on the neurohormone profile of 100 patients, has shown the effectiveness of our method in 70% of the cases.[45]

The barbiturate enzyme induction could not hitherto be put to practical use because of the sedative effect of the barbiturates. The advent of proxibarbal has solved this dilemma. The next step will be the specific study of the induced enzymes in vitro and their relationship to cytochrome P-450. The use of proxibarbal for the treatment of suitable cases of weather-sensitivity, migraine, serotonin abortion, sprue, intractable histamine allergy, incurable skin and nervous diseases, and hyperthyroidism is obvious.[46]

The unknown factors involved in enzyme induction with proxibarbal are the following:

1.   Only up to 70% of patients react to it; the other 30% react too slowly, though they do not show any adverse effects.

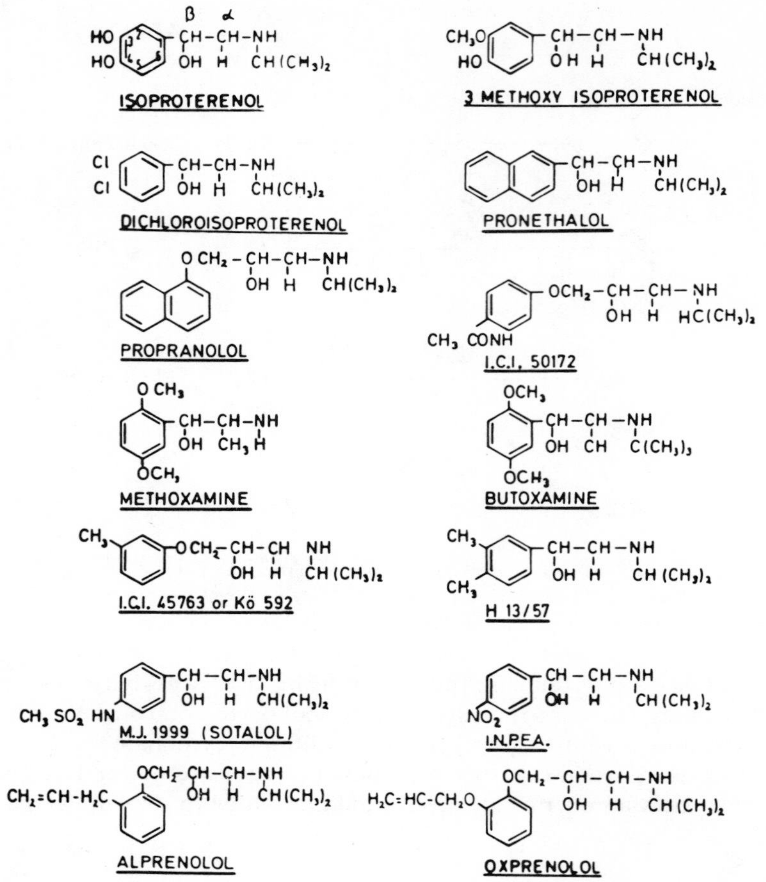

FIGURE 17.   Beta-adrenergic drugs, called for short "beta-blockers", are drugs which block the beta-receptors for epinephrine. As they regulate pulse frequency and blood pressure, they are mostly used for slowing-down of tachycardia (heart acceleration) and heart palpitations in hyperthyroidism. The most popular beta-blockers are propranolol (Inderal® — ICI), sotalol (Sotacor — Bristol), alprenol (Corbetan — Teva), oxprenolol (Trasicor — Ciba). Recently, Pindolol (Visken — Sandoz) has been added to the list.

2.   The pathways of the reaction are not yet known; however, experience with proxibarbal in thousands of patients suffering from migraine has shown it to be innocuous.
3.   The initial dose for enzyme induction was routinely fixed by us at 3 × 100 mg/day for the first 4 to 8 weeks; subsequently a maintenance dose of 100 mg 1 to 3 times weekly seems to be sufficient, but this needs further study.

Enzyme induction — considered since its discovery to be a disadvantage of barbiturate administration[47] — may now become a major tool in fighting enzymatic psychosomatic and metabolic reactions to climate changes resistant to orthodox therapy.

5. Drugs Against Heat
Tropical and subtropical zones require lower drug dosages than the temperate zones. This has been our daily experience and has induced us to start every treatment with quarter doses of drugs. Still, there are exceptions to this rule and antibiotics should

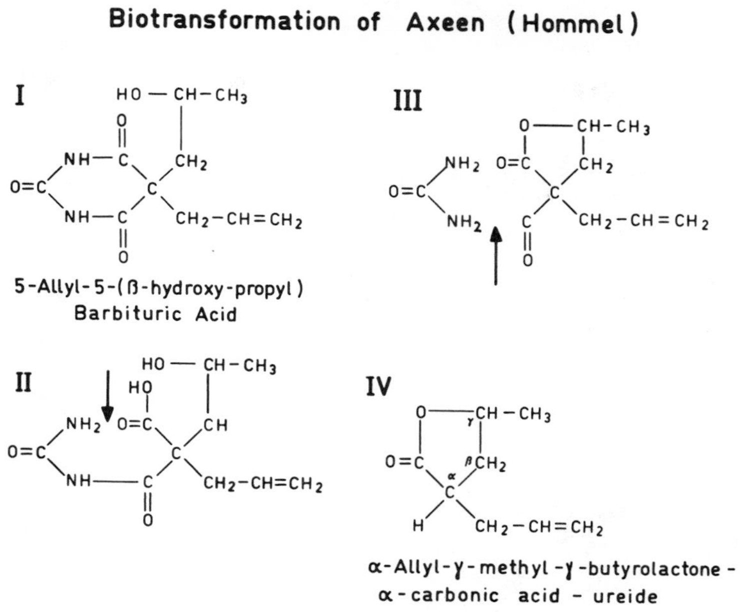

Biotransformation of Axeen (Hommel)

I

5-Allyl-5-(β-hydroxy-propyl)
Barbituric Acid

III

II

IV

α-Allyl-γ-methyl-γ-butyrolactone -
α-carbonic acid - ureide

FIGURE 18.   Proxibarbal (Axeen — Hommel) is a barbiturate without sedative effect which is metabolized to a 5-membered lactone. Its only barbituric property is its ability to induce enzymes which destroy a surplus of neurohormones and rid people of their allergy and their neurovegetative sufferings, including migraine. Its biotransformation runs through four stages: (I) Proxibarbal; (II) allophanyl-2-allyl-4-valerolactone; (III) urea + alpha-allyl-gamma-methyl-gamma-butyrolactone-alpha carbonic acid; (IV) alpha-allyl-gamma-methyl-butyrolactone.

surely be given at full dosage schedules to avoid the emergence of drug-resistant bacterial strains. Work on laboratory animals conducted by Weihe[48] has also shown that their requirement of drugs decreases with increasing temperature.

We have exposed young rats to temperatures increasing from 23 to 37°C. At 37°C the animals could not survive for longer than 5 days, as they died of heat stroke due to thyroid hyperactivity. At this stage they were given drugs to let them overcome the future heat stress.[49,49a] Table 8 shows the 20 drugs and hormones tested and, amazingly, the two hormones listed at the end of the table allowed all rats to survive for unlimited periods. These two miracle drugs are rarely used in medicine; one is dehydroepiandrosterone (Diandron-Organon®) used as a roburant; the other is dihydrotachysterol AT-10 also called *Hytakerol®* (Winthrop), which is used in severe cases of calcium deficiency conducive to tetanic cramps (AT-10 means anti-tetany). It is obvious that with the aid of such drugs man could overcome high temperatures in tropical countries, yet their practical use is still in an investigational stage.

D. Air Ionization
1. Beneficial Effect

Negative ionization for therapeutic purposes is derived from ionizing sources, e.g., an electric transformer of 5 to 10 kV with a needle at its end which emits up to 50,000 times more negative than positive ions (Figure 19). This is the so-called corona discharge; it produces a strong ozone current which disappears at a distance of 1 cm.

Negative ions abolish the adverse effects of the positive ions. However, not everybody enjoys negative ionization — only 30% of the population need negative ioniza-

Table 8
EFFECT OF DRUGS ON AMBIENT HEAT STRESS — SURVIVAL OF IMMATURE FEMALE RATS AT 37—38°C AND 40% RELATIVE HUMIDITY

No.	Drug group	Generic name	Proprietary name	Possible mechanism of action	Dose (mg/kg/day)	Survival rate (%)
1	Control	0.9% NaCl	Normal saline	—	0.2	0
2	Metabolic coenzyme	Glutamyl-cysteinyl-glycine	Glutathione	Metabolic changes	10—100	0
3	Barbiturate	Phenobarbitone	Luminal	Enzymatic changes	1—10	0
4	Antipyretic	Acetaminophen	Paracetamol	Heat center inhibition	1—5	20
5	Serotonin antagonists	Pizotyline	Sandomigran		0.01—0.1	0
6		Methysergide	Deseril, Sansert®	Heat	0.01—0.1	0
7		Methergoline	Lyserdol	Center	0.01—0.1	0
8		Cyproheptadine	Periactin®	Mediator block	0.01—0.1	0
9		D-2-bromo-lysergic diethylamide	BOL-148ᵃ		0.01—0.1	40
10	Phenothiazine	Mepazine	Pacatol	Heat center level lowered	0.01—0.1	50
11	Gonadotropic hormone	HCGᵇ	Pregnyl	Progesterone	10—100 U	0
12	Progestagen	Progesterone	Progestin	Hyperthermia	0.5—5	0
13	Retroprogesterone	Dydrogesterone	Duphaston	Hypothermia	0.5—5	0
14	Estrogen	Estriol	Ovestin		0.01	0
15		Estradiol	Dimenformon	Catabolism	0.001	0
16	Androgen	Nandrolone Phenpropionate	Durabolin		1—2	0
17		Methandriol	Neosteron	Anabolism	2—20	0
18		Testosterone Phenpropionate	TPPᶜ		2—20	0
19	Hypercalcemic	Dehydroepiandrosterone	Diandron		5—50	100
20		Dihydrotachysterol	AT-10,ᵃ Hytakerol®	Thyroid block	0.1—0.5	100

Note: Groups of 6 rats per doses of survival drug; period of treatment: 21st to 42nd day of age; food and water intake: *ad libitum*; initial body weight 40 ± 5 g; final body weight 70 ± 10 g; drug and saline control administered daily s.c. in 0.2 mℓ.

[a]   BOL-148 = D-2 bromolysergic acid diethylamide.
[b]   HCG = Human chorionic gonadotropin.
[c]   TPP = Testosterone phenpropionate.
[d]   AT-10 = Anti-Tetanin-10.

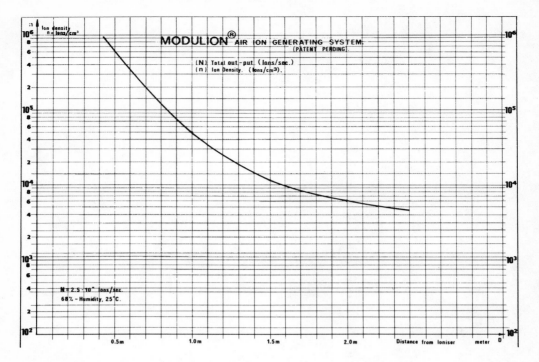

FIGURE 19.   Diagram of negative ion output from an ionizing apparatus (Modulion®). At 0.5 m distance it provides 5×10^6 negative ions/cm³/sec; at 2 m, which is the optimal distance for a patient, 5×10^4 negative ions/cm³/sec are available.

tion, and another 30% feel well if the amount of negative ionization in the air is increased, leaving 40% who are indifferent to ionization. However, this is reason enough to recommend the ionizing apparatuses. A patient who does not need them will not suffer any damage from using them and he simply ignores them. The patient who does need them will feel the benefit within a few minutes and report an easing of the various above-mentioned complaints from serotonin.

For this reason all the manufacturers of ionizing apparatuses have made provision for the purchaser to try out the apparatus for a period of 2 to 4 weeks before deciding whether to buy it or not. The cost of running such an apparatus is very low (despite the high voltage), amounting up to 25 to 40 W, which corresponds to that of a small lamp.

2. Adverse Effects

Noxious effects of negative ionization have never been described, and we have shown that they do not exist.[50] The reason for this is that negative ions decay within a few seconds. This is also why ionizing apparatuses can be used with the same beneficial effect with the windows open or closed. The ionizing apparatuses marketed in Europe do not entail any danger of overdosage because their activity is more or less uniform.

One cannot preserve ions in one's room and to try to do so would be to follow in the footsteps of those wise people who forgot to put windows in their houses and then tried to carry the sunshine indoors in sacks. For the same reason there is little danger of a room receiving an overdose of negative ions from ionizing apparatuses.

Adverse aftereffects have not been noted and it can hardly be assumed from the medical point of view that these could exist. The apparatuses on the market can ionize a room of up to 5×5 m and provide relief for a patient within a radius of 2 m. Still, even here there are no rules, since many patients prefer a 1 m distance while others prefer 3 m. This means that the patient can regulate the ion concentration and uptake by varying his distance from the apparatus.

3. Practical Application

The number of ions emanating from ionizing apparatuses is between 50 and 500,000 negative ions per cubic centimeter at a distance of 1 m. At 2 m distance the ions would fluctuate between 20 and 100,000/cm³. Higher values are not needed because they would only create an ozone surplus. The optimal range for a patient is 2 to 4 m distance from the apparatus (Figure 19).

Therapeutic use needs no special arrangement: the patient can switch on the apparatus at night if he is awakened by his complaints. He may run the apparatus all night and even permanently; closing of windows is unnecessary. In work rooms there is no objection to having the apparatus running during the 8-hr working day with the windows either open or closed, as desired.

The combined use of ionizing apparatuses with humidifiers or air cleaners is unnecessary, though desirable. A combination with air conditioning is certainly possible but needs ion control to be sure that the air conditioning does not affect the ionizing effect. The combination of an ionizing apparatus with a DC electrofield generator has recently been recommended by us.[51]

E. Air Conditioning

Air conditioning can be defined as the control of the temperature, moisture content, and purity of the air in a room or building in order to create a comfortable environment for the occupants. A most effective and inexpensive method of cooling a house is the ''desert cooler'' of Africa. It lets water drop outside the window through a straw or plastic screen and creates evaporation cold and negative ionization. The high effectiveness of this simple device is astonishing.

The main principle in modern air conditioning is the passing of an air current at normal or preheated temperatures through a humidifying chamber in which a fine water spray is discharged. Soluble gases and suspended particles are removed and a clean moist air current leaves the chamber. A thermostatic device controls the temperature at which the air leaves the spray chamber and also its water vapor content. In order to reduce the amount of water vapor, the air should be cooled below its dew point. The reduction of humidity facilitates sweat evaporation during high temperatures and increases our comfort in humid tropical climates considerably.

The object of air conditioning is to make the air inside a building more suitable for the purpose for which the interior is to be used. Full air conditioning implies full control of the temperature of the air, its moisture content, filtration of suspended particles, and sometimes covers chemical washing to remove gaseous contaminants. In some buildings there is no recirculation of air, the conditioned air being exhausted directly to the exterior and being entirely replaced by fresh air from outside. This is healthy, but expensive, and it seems more economical to recirculate a large proportion of the conditioned air, making up one fifth with fresh air. Yet, this system has its drawbacks as we shall see immediately.

Two important ecological problems arise with reconditioned air conditioning. One is the recirculation of bacteria and viruses with other internal atmospheric pollutants like tobacco smoke from room to room down the ducts. This may, for example, present problems of cross-infection in hospital practice, particularly if recirculated air is used; a good example was the spread of Legionnaire's disease in a Philadelphia hotel. The other is the problem of duct noise. In badly designed systems, noise may pass from room to room or along from the plant.

There is no doubt that air conditioning is the most important invention to maintain fruitful work and mental capacity under adverse conditions of climate change. Yet, the exclusion of fresh air current, sunlight, and air ionization is a severe drawback which can only be resolved by exposing a person to air conditioning only for part of the day. The danger of positive ionization by air conditioning can be neutralized by negative ionizers.

REFERENCES

1. Sulman, F. G., *The Effect of Air Ionization, Electric Fields, Atmospherics and other Electric Phenomena on Man and Animal,* Monogr., Charles C Thomas, Springfield, Ill., 1980.
2. Sulman, F. G., *Health, Weather and Climate,* Monogr., S. Karger, Basel, 1976.
3. Selye, H., *Hormones and Resistance,* Springer, Berlin, 1971.
4. Loewi, O., Ueber humorale Uebertragbarkeit der Herznerven Wirkung, *Pfluegers Arch. ges Physiologie,* 189, 239, 1921.
5. Curry, M., *Bioklimatik,* Riederau-Ammersee, American Bioclimatology Research Institute, West Germany, 1946.
6. Sulman, F. G., Pfeifer, Y., Levy, D., Lunkan, L., and Superstine, E., Human weather sensitivity and atmospheric electricity, in *Israel Meteorological Research Papers,* Steinitz Memorial Volume No. 1, 1977, 42.
7. Tal, E. and Sulman, F. G., Urinary thyroxine test, *Lancet,* 1, 1291, 1972.
8. Sulman, F. G., Tal, E., Pfeifer, Y., and Superstine, E., Intermittent hyperthyreosis — a heat stress syndrome, *Hormon. Metabol. Res.,* 7, 424, 1975.
9. Sulman, F. G. and Tal, E., Treatment of functional hypothyroidism by oral TRF monitored by daily urinary thyroxine and histamine assay, *Hormone Metabol. Res.,* 6, 92, 1974.
10. Dikstein, S., Kaplanski, Y., Koch, Y., and Sulman F. G., The effect of heat stress on body development of rats, *Life Sci.,* 9, 1191, 1970.
10a. Dikstein, S. and Sulman, F. G., Prevention of cold stress by anabolic agents, *Isr. J. Med. Sci.,* 8, 572, 1972.
11. Faust, V., *Biometeorologie,* Hippokrates, Stuttgart, 1977.
12. Behar, A. J., Deutsch, E., Pomerantz, E., Pfeifer, Y., and Sulman, F. G., Migraine, serotonin and the carotid body, *Lancet* 1, 550, 1979.
13. Rim, Y., Psychological test performance during climatic heat stress from desert winds, *Int. J. Biometeor.,* 19, 37, 1975.
14. Rim, Y., Psychological test performance of different personality types on sharav days in artificial air ionisation, *Int. J. Biometeorol.,* 21, 337, 1977.
15. Charry, J. M., Behavior of man and animals under different conditions of air ionisation, personal communication, 1978.
16. Assael, M., Pfeifer, Y., Sulman, F. G., Alpern, S., and Shalita, B., Influence of artificial air ionisation on the human electroencephalogram, *Int. J. Biometeorol.,* 18, 306, 1974.
17. Sardi, Z., Effect of negative air ionization on psychotechnical tests, Hebrew University-Hadassah Psycho-technical Services for Professional Advice, Jerusalem, personal communication, 1977.
18. Rheinstein, J., The influence of artifically generated atmospheric ions on simple reaction time and on the optical moment, Dr. Dissertation, Techn. Hochschule Munich, 1960.
19. Kerdoe, I., Hay, G., and Srab, F., New possibilities in the increasing of driving safety, *Medicor News Budapest,* 4, 17, 1970.
20. Hay, G., Mészáros, I., Pelyhe, H., and Srab, F., New tests to prove the favourable effect of the miniature ionizer in driving, *Medicor News Budapest,* 3, 14, 1972.
21. Soyka, F. and Edmonds, A., *The Ion Effect,* E. P. Dutton & Co., New York, 1977.
22. King, G. W. K., *Annual Bibliographies on Ionization of the Air,* American Institute Med. Climatology, Philadelphia, 1960-1975.
23. Nir, I. and Sulman, F. G., Effect of chronic negative air ionization on the mice and rat estrous cycle, Unpublished results, 1977.
24. Sulman, F. G., Pfeifer, Y., Weinstein, D., Sadovsky, E., and Polishuk, W. Z., Serotonin as a mediator of psychosomatic disturbances in fertility and sterility and its elimination by ionizing treatment, in *4th Int. Congr. Psychosom. Obst. Gyn., Tel Aviv,* S. Karger, Basel, 1975, 273.

25. **Lang, S. and Lehmair, M.**, Bioklimatische Wirkungen der elektrischen Umwelt am Arbeitsplatz, *Dtsch. Architektenblatt,* 3, 209, 1977.
26. **Koenig, H. L. and Lang, S.**, *Unsichtbare Umwelt,* 2nd ed., Technical University, Munich, 1977.
27. **Landau, S. F. and Drapkin, I.**, Ethnic Patterns of Criminal Homicide in Israel, *Ford Foundation Rep.* 5 F-8, Hebrew University, Jerusalem, 1968.
28. **Kuhnke, W.**, Bericht ueber bisherige medizin-meteorologische Erfahrungen füer Nordwest-Deutschland, Meterological Institute, Hamburg, 1960.
29. **Sulman, F. G., Levy, D., and Lunkan, L.**, Wetterfuehligkeit und ihre Beziehung zu Sferics, Ionen und Electrofeldern, *Zschr. f. Physik. Medizin,* 5, 229, 1976.
30. **Sulman, F. G., Levy, D., Lunkan, L., Pfeifer, Y., and Tal, E.**, Human reactions to climatic fluctuations, in *Int. Conf. Meteorol. of Semi-Arid Zones,* Government Meteorological Service, Tel Aviv, 1977, 54.
31. **Rehn, E.**, *Die Blutgerinnung,* Spitzner, Ettlingen, Germany, 1970.
32. **Merimsky, E., Litmanovitch, Y. I., and Sulman, F. G.**, Prevention of post-operative thromboembolism by negative air ionization in a double-blind study, *Proc. 9th Int. Congr. Biometeorol.,* Osnabrueck University, West Germany, 1981, 1.
33. **Clawson, M.**, The influence of weather on outdoor recreation, in *Univ. Chicago Dept. Geogr. Res. Rep. 105,* Sewell, W. R. D., Ed., University of Chicago, 1966, 183.
34. **Harlfinger, O.**, Vacation guide, *Arch. Met. Geophys. Bioklim. Ser. B,* 23, 81, 1975.
35. **Jokl, E. and Jokl, P.**, *Exercise and Altitude,* S. Karger, Basel, 1968, 199.
36. **Minkh, A. A.**, Aero-Ionization in Medicine, Vis. Joint Publ. Res. Service, Washington, D.C., 1961, 1.
37. **Rivolier, J. R., Herisson, Y., and Zouloumian, P.**, Research on a possible action of negative over-ionization on healthy humans. I, *Biometeorology,* 6(Suppl. 18), 134, 1975.
38. **Sulman, F. G., Pfeifer, Y., and Superstine, E.**, The adrenal exhaustion syndrome: an adrenal deficiency in long-distance runners, *Ann. N.Y. Acad. Sciences,* 301, 918, 1977.
39. **Miller, W. H.**, Santa Ana winds and crime, *Prof. Geog.,* 20, 23, 1968.
40. **Richner, H.**, *Zusammenhaenge zwischen raschen atmosphaerischen Druckschwankungen, Wetterlage und subjektivem Befinden,* Lab. f. Atmosphaerenphysik ETH, Zurich, 1974.
41. **Ungeheuer, H.**, Vom Foehn des Foehnkranken, *Fortschr. Med.,* 74, 357, 1956.
42. **Brezowsky, H.**, Morbidity and weather, in *Medical Climatology,* Licht, S., Ed., Elizabeth Licht, New Haven, 1964, 358.
43. **Posse, P., Kleinschmidt, J., Pratzel, H., and Grohmann, K.**, Laengsschnittuntersuchungen zur Wetterwirkung auf die Ausscheidung von Serotonin und 5-Hydroxy-indolylessigsaeure in Urin., *Z. Physikal Med.,* 6, 67, 1977.
44. **Sulman, F. G.**, Prevention of allergy by enzyme induction, *Lancet,* 1, 1206, 1977.
45. **Sulman, F. G., Pfeifer, Y., and Goldgraber, M. B.**, Treatment of multiple allergy by enzyme induction with proxibarbal., *10th Int. Congr. Allergology,* Jerusalem, 1979, 316.
46. **Sulman, F. G.**, Weather sensitivity, its diagnosis and treatment, *Hexagon Suppl. (Basel),* 5, 1, 1977.
47. **Conney, A. H.**, Pharmacological implications of microsomal enzyme induction, *Pharmacol. Rev.,* 19, 317, 1967.
48. **Weihe, W. H.**, Meteorological effects in laboratory animals, in *Progress in Biometeorology,* Vol. 3, Tromp, S. W., Ed., Swets & Zeitlinger, Amsterdam, 1976, 119.
49. **Tal, E. and Sulman, F. G.**, Rat thyrotrophin levels during heat stress and stimulation by thyrotrophin releasing factor, *J. Endocrinol.,* 57, 181, 1973.
49a. **Tal, E. and Sulman, F. G.**, Dehydroepiandrosterone-induced thyrotrophin release in immature rats, *J. Endocrinol.,* 57, 183, 1973.
50. **Sulman, F. G., Levy, D., Lunkan, L., Pfeifer, Y., and Tal, E.**, Absence of harmful effects of protracted negative air ionisation, *Int. J. Biometerol.,* 22, 53, 1978.
51. **Sulman, F. G.**, Physikalische Therapie der Wetterleiden, *Phys. Med. Rehabil.,* 21, 220, 1980.

Chapter 2

ECOLOGICAL IMPACT OF SHORT-TERM CLIMATE CHANGES

I. ENVIRONMENTAL PROBLEMS

A. Agriculture
1. Effect of Humidity

The effects of weather variations on agriculture are often far-reaching. Generally, only the direct effect of weather and climate on the yield and the quality of the specific crops is considered, but any analysis of the total crop industry shows that the influence of the climate does not end on the fields of the farm.

It is essential that more research along these lines be done, in view of the fact, as stressed by Watson,[1] that "climate determines the crops the farmers can grow, weather influences the annual yield and hence the farmers' profit, and more important, especially in underdeveloped countries, how much food there is to eat."

Associated with the supply of water for agriculture are the properties of evaporation and transpiration, and the concept of potential evapotranspiration, all of these factors being especially important in the assessment of the water needs of agriculture. Scientific irrigation will accrue not only to the irrigation farmer through increased yields and profits but also to his neighbor who turns to irrigation, and to those who provide various services to the farmer, such as the manufacturers and dealers of irrigation equipment, the electric power companies, the fertilizer industry, and the general business community.

In old Egyptian times the amount of water made available to the inhabitants by the Nile inundations in July was measured by the nilometers. Their height indicated how much crop the peasants had to deliver to the Pharaoh or the tax to be paid on it.

2. Effect of Technology

Much of his lift in per acre yields is due to the effort of man — what he does with his land, his seeds, his fertilizer, his pest and disease control, his row spacings — but much more may be the result of a fundamental change in weather and climate. There has been a growing tendency to believe that technology has reduced the influence of weather on grain production so that we no longer need fear shortages due to unfavorable weather. However, there is increasing evidence[2] that a period of favorable weather interacted with technology to produce our recent high yields, and that perhaps half of the increase in yield per acre since 1950 has been due to a change to more favorable weather for grain crops.

The significance of agroclimatological studies is not lessened by advances in technology. Indeed, increasing pressure of populations on world food supplies suggests that such studies will be of even more value in the future than they are today. An integral part of developing agricultural technology will be based on rain seeding which has not yet been harnessed enough for increasing crop yield.

3. Effect of Ionization

Recently a new solution has been proposed for the increase of crop production under intensive treatment. The yield of tomatoes, cucumbers, and similar produce in conservatory greenhouses can be doubled by ionization with 10,000 to 100,000 ions per cubic centimeter of air. The method which is neither expensive nor dangerous promises much improvement of crop yield wherever greenhouse growing is applicable.[3]

The idea of using artificial overhead discharge for electroculture of plants was inaugurated by Lemstroem[4] in 1883 and published in 1904. Sidaway[5] reported that use of a high tension system of 1.25 m mesh-wire networks supported about 0.4 m above the crop and provided with downwardly directed discharge points resembling the "barbs" of barbed wire, has increased crop yield by 20%. The method was successfully applied to growing cucumbers, strawberries, wheat, oats, and barley.

B. Forestry

1. Impact of Forests on Climate

Several aspects of forests and forestry are of importance in a study of the value of the weather. Forest canopies, for example, influence the hydrology of different areas by reducing the amount of precipitation that reaches the ground, and by decreasing the amount of evaporation from the ground. A forested area also acts as a natural environment for wildlife, and in many localities provides income through tourism, meat, and timber. In addition, the influence of weather and climate on forests, the influence of forests on the local climatic conditions, and the importance of both man and nature with respect to forest fires are significant facets of weather and forestry.

The relationships between the local climatic conditions and the local vegetation, whether natural or man-planted, are of considerable interest to biogeographers. However, the effect of vegetation, including forests, on the local climatic character is probably of greater significance in any assessment of the value of the weather. In addition to the effect that forests and shelter belts have on the local climate, and the relationships that exist between climatic conditions and "natural" vegetation, there is a clear dependence between tree growth and climate. One evidence of this is seen in tree rings, which have been studied in considerable detail all over the world as a means of paleoclimatologic dating (see Chapter 3).

2. Forest Fires

In many areas of the world the traveler is confronted with signs "Prevent Forest Fires", based on the suggestion put forward by state forest services that "Only You Can Prevent Forest Fires". But in some areas this is far from the truth, for lightning and not man is often the major cause of forest fires. A further aspect of weather and forest fires are those atmospheric conditions which create fire hazard states, notably low humidity, hot drying winds, low rainfall, and sunshine. Weather and climate are major variables in the behavior of all outdoor fires, and in the forest, unanticipated changes in the weather can turn a moderate wildfire into a holocaust.

An associated feature of forest fires is the effect of smoke from the forest blazes on visibility and the radiation balance, which can have significant effects on climate. For example, smoke from forest fires, if moved by winds over an urban area, may reduce the normal solar radiation by as much as 30 to 60%. This reduced sunlight may in turn cause lowered visibility and crop yield, increased use of power for artificial illumination, and probable delays in land and air transportation.

Forest fires, whether started by man or lightning, create significant economic and social problems. For example, such fires cause damage to watersheds, wildlife, grasslands, and outdoor recreation facilities. In addition, fires can cause loss of life and the destruction of homes and other property. Moreover, the supply of forest products and the economic well-being of many forest-resource based industries are often adversely affected by fires.

The well-known effect of fire to create positive air ionization is another noxious factor which makes forest fires highly undesirable and calls for thorough prevention.

3. The Sequoia

The importance of forest fires as an ecological factor is well demonstrated by the

Sequoia forests of California. Sequoias have been menaced for thousands of years by lightning fires. The result was that only the fittest survived and their power of resistance was inherited by the "offspring". Visiting Sequoia forests one is often surprised to see "senile" but flourishing trees split by lightning. Thus it seems that the Sequoia is a good lightning conductor.

Many people remember the remarkable story of the "Mother of the Forest", the pride of the Calaveras Grove in California. This tree was until 1854 one of the most beautiful Sequoias, reaching up to 170 m height. When its bark vanished up to a height of 38 m, it was destroyed by a lightning stroke which burned the unprotected lower stem of the giant. It is conceivable that the wet bark of such a tree, being 60-cm thick, protects a Sequoia against lightning strokes and establishes its ability as a lightning conductor. For this reason poor "Mother of the Forest" had to die. Counting of the wood rings yielded 4000 annual rings corresponding to an age of 4000 years at least. Such ages have been corroborated by the findings of fossil Sequoias all over the world.

C. Fishing

1. Fishing Problems

A popular sport and industry that is weather-sensitive and at the same time economically important in many countries is fishing. The direct effect of adverse weather conditions on the fishing industry is obvious, and it is common knowledge that, during adverse weather conditions, many fishing boats are either unable to leave port or are forced to return to port earlier than anticipated. The weather-related factors in turn affect the retail price of fish, and if adverse weather conditions continue for any length of time, the decreased catch of fish would have significant economic effects on the fishing community.

There is no proof that fishing is influenced by air electricity, yet it is an accepted fact that on days of increased air electricity the catch of fish is reduced. The significance, however, of this observation is questionable since high air electricity is always connected with adverse weather conditions, as described in the following paragraph.

2. Temperature Problems

Another weather-related factor is the effect of colder than normal, or warmer than normal, ocean temperatures on the total fish population and hence the probable catch. It may happen that the extremely cold or unusually warm air temperatures during 1 month or over a limited period of the year are reflected in midwater or bottom temperatures in the oceanic regions where halibut spawn and where the eggs and larvae appear to spend their early existence.

3. Angling

Suggestive correlations exist between ambient temperature and "catch" of some inland species, so that weather relationships cannot be ignored. One important difficulty in establishing such correlations is the measurement of the "catch", for unlike a wheat field with its unique farmer, rivers and lakes are open to all kinds of fishermen. Thus the catch data on which some climate-catch evaluations are based may not necessarily reflect the total catch of the area. In addition, it should be noted that the effect of weather on fishing does not necessarily apply only to the commercial fishing industry, for, as any angler will testify, under the right kind of weather conditions the fish really do bite better!

An innovation in climatology was heralded by the introduction of fish ponds. In Israel this was once considered taboo because of the danger of spreading malaria. Practice showed that the fish, especially carp and St. Peter's fish, devoured malaria larvae. Moreover, the presence of many lakes changes the climate completely.

D. Animal Breeding

1. Pastoral Production

In some parts of the world agriculture based on grassland farming is extremely important. Moreover, if one considers the pastoral products of milk, butter, cheese, meat and wool, and their world distribution in terms of exports, then several countries, notably Australia, New Zealand, South Africa, and Argentina, are especially important, particularly in view of the fact that their climatic year is "opposite" to that of the major importing areas of the world (U.S. and Europe).

Perhaps the major climatic problem facing annual production, however, is the effect of drought, which is a serious problem to many aspects of successful meat production. Thus, e.g., in Australia, the influence of drought conditions on beef and cattle numbers is of enormous importance. The duration and distribution of rainfall, temperature, and evaporation can exert a tremendous influence on the structure of the meat industry and dairy husbandry, all of them having a practical impact on climate control.

2. Sheep Production

There are many reasons for the differences in wool clips, including the time at which the sheep are shorn, but clearly climatic variations are among the major aspects to be assessed, and some of the influences of climate on wool growth in several countries are of practical importance. The impact of rainfall on wool growth has been reported from studies in Scotland, Argentina, and Australia where most work has shown the general benefit of rainfall. For example, the summer and autumn rainfall has the greatest influence on fleece weight in the following season and winter rainfall is the least influential. Of greater importance is the effect of drought conditions on wool production, which would imply that the problems of effective drought control constitute one of the greatest challenges to efficient and increasing sheep production.

It is clear that the relationship between wool quantity, wool quality, and climatic conditions is not a simple one. The most extensive and more readily apparent of the changes in the environment from one season to the next are those that come about as the result of variations in the weather. Some of the observed wool clip variations are due to the time of shearing, yet differences may be seasonal in character and dependent mainly on ambient temperatures and the availability of feed in the winter and early spring months.

3. Dairy Husbandry

Cow breeding may present fewer problems since intensive cow husbandry is now carried out in cow sheds. Still, the influence of the weather on milk "letting-down" of cows is well known to every cow breeder and attendant. Modern methods try to overcome any detrimental effect of the weather by introducing rotating milk podiums (parlors) where prolactin release by a relaxed atmosphere, conditioned reflexes, and feeding proteins during milking increase milk yield.[6] Prolactin release is promoted by serotonin and suppressed by epinephrine. Thus it is positive air ionization which can increase milk yield by serotonin release; and it is bad handling of the cows which could lower milk yield by epinephrine release.

E. Urbanization

1. Urban Heat

Since a large proportion of the world's population, particularly in North America, Europe, and Australia, lives in cities, the climate of urban areas is of importance to many people.

The characteristic warmth of a city is called the urban heat island. It is at a maximum at night when skies are clear and winds are light, and its seasonal maximum is summer to early fall. A beneficial effect of the nighttime heat island is to lengthen the frost-free period of the year. Daytime city temperatures are usually about the same as, or even slightly lower than, those in the suburbs.

2. Urban Climate

A city is a collection of microclimates, the character of the built-up area immediately surrounding the center probably being more important than the size or the form of the city. In addition to the heat island effect, downtown areas of cities also have fewer natural sources of evaporation and transpiration. The water vapor content of city air is therefore less in summer, as evidenced by lower frequencies of dense fog, than in nearby countryside. By contrast, in winter the situation is often reversed because of the emission of water by combustion processes.

The city's effect on its own climate is, therefore, complex and far-reaching. Landsberg[7] concluded that cities in the middle latitudes receive 15% less sunshine on horizontal surfaces than is received in surrounding rural areas, and that they receive 5% less ultraviolet radiation in summer and 30% less in winter. Landsberg's figures also show that the city, compared with the countryside, has a 6% lower annual mean relative humidity, 5 to 10% more precipitation, 5 to 10% more cloudiness, 25% lower mean annual wind speed, 30% more fog in summer, and 100% more fog in winter.

The advantages and the disadvantages of city climate testify to the fact that the city's climate is distinctly different from that of the countryside, and a fuller understanding of the climatic changes created by a city may make it possible to manage city growth in such a way that the effect of troublesome atmospheric changes will be minimal. A knowledge of the climate of cities has many important applications, notably in bioclimatology, industrial climatology, and economic climatology. In particular, in cities where pollution levels are becoming alarmingly high, it will become increasingly necessary to select new areas for industrial zoning on the basis of meteorological factors, as well as taking into account variations in the modified urban weather in transportation planning.

In this conglomeration of facts air electricity plays an important role. Air ions affect human well-being, sferics are considerably suppressed by the accumulation of housing, and oscillations of electrofields depend very much on the space left for them by crowding of houses. These oscillations, if abundant, tend to create a euphoric mood in city inhabitants.

3. Town Planning

Consideration of the effect of climate on town planning involves two factors, first in the initial siting of new towns or new industries, and second in the location of industries or new settlements as part of already established areas. In the first case, the initial siting of a new town or industrial complex should be planned with knowledge of the climatological conditions, and some of the "new" towns that have been developed in various parts of the world in the 20th century have in some cases utilized climatological information above all other data in their planning. However, most towns and cities are located where they are today because of historical, political, geographical, or economic reasons, and the climatologist has had little say, if any, in choosing the site.

Planning, however, does not end once the site has been chosen, and it is in the area of urban growth and development that weather and climatic information have their most immediate value. For example, the siting of the residential sections of an urban area on the windward side, of industrial establishments on the usual lee side of a development, and the benefits of parks for recreation have long been recognized. In addition, the planning of street widths and street siting to take account of the prevailing cold and warm wind directions is highly desirable. Grassy and tree areas are also important in decreasing the dust content of cities, and in decreasing the maximum temperatures in their vicinity. Planning based on applied climatological data can also aid in the size and spacing of buildings. In cold climates, for example, it is highly desirable from a fuel consumption viewpoint that houses be built closely together.

Irrespective of the planning that is so necessary for the successful development of urban and industrial areas, there remains the important factor that in any urban area new microclimates are constantly being created, with consequent alterations in air humidity and heat balance. One facet of the microclimates so formed is atmospheric pollution, a factor which must be taken into account in both short- and long-term planning if man is to live in harmony with his urban environment. Indeed, it is clear that in many areas the "value" of the weather, as it is associated with industry and air pollution, is becoming a vital factor in the planning of our cities. Moreover, it is also clear that successful planning will only be accomplished if it is done with proper consideration for the atmosphere, as well as for the economic and social environment.

4. Leonardo's Città Ideale (the "Satellite Town of the Renaissance")[7,7a]

After the severe plague epidemic of 1484 to 1485 in Milan, Leonardo designed plans for an ideal town with straight, wide streets with drains. These streets were intended for pedestrians only with a corresponding mirror image of subterranean roads for the entire goods traffic and transport of waste. Leonardo's design, thought out in great detail, shows a fascinating boldness in the removal of waste and in traffic regulation (already anticipating the one-way traffic principle, e.g., in the separate entrances and exits of the communicating stairways) and his ideas on the relieving of the overcrowded towns by creating satellite settlements on the periphery.

Leonardo da Vinci (1452-1519) has bequeathed to mankind a legacy of some 7000 designs on mechanics, aviation, anatomy, etc. Those devoted to urban planning are still of practical interest. His town planning pictures are also of high aesthetic and hygienic value. He designed them for Lodovico Sforza, his patron, who, ruling Milan in 1451 to 1508, recognized that the ravages of a plague epidemic which destroyed one third of his population required an urgent revision.[8]

Leonardo proposed the removal of the poverty-stricken population (Poveraglia) to ten satellite towns on the periphery of Milan as the best remedy for pestilence, overcrowding, and rowdyism. He advocated building of satellite towns near big rivers — not foreseeing the modern dangers of pollution — but keeping them clean by dams, drains, and manholes. He suggested building of houses at distances from each other to create gardens, greenery, and recreation sites. This would also lessen the dangers from spreading fires and epidemics.

Streets should be as broad as the adjacent quarters are high and he discouraged "skyscrapers" as they would create too much shadow: "Dove non viene il sole — viene il medico" — where the sun does not come — the physician comes.

His most ingenious proposal was the structure of towns on three levels (Figure 1). The upper level for living and balconies, the middle level for shopping arcades under the balconies, the lower level for traffic and canalization. Communication between the three levels would be by helix-shaped spiral staircases or double-staircases (Figures 2 and 3).

Toilets were specially designed to remove wastewater and stench by rotating doors (tornos). Horse stables on the lower level had special designs for feeding the animals and removing their excrements (Figure 4). Thus, every item of urbanization was meticulously planned. Special designs were provided for the palaces of the noblemen (Figure 5), and even a proposal was found how to build a brothel where the cavaliers would never meet each other.

FIGURE 1.   Leonardo da Vinci's "Città ideale" is the satellite city of the Renaissance. It has its traffic lanes on three levels: left: living quarter, right: traffic quarter with an upper story designated for pedestrians and shopping and a lower story for horses, vehicles, and sewers. Note the mirror writing preferred by Leonardo for important remarks.

Leonardo's ideas have been successfully applied to modern town planning, as will be shown in the next section on Landsberg's Metutopia.

5. The Ideal Urbanization — Metutopia

Landsberg[9] has conceived the concept of modern town planning called by him *Metutopia.*[10]

In the utopian town — let us call it *Metutopia* — we would postulate preservation of as many natural trees as possible. Those that succumb would be replaced and wherever feasible new trees would be added. The summer heat-island problem suggests that in our utopian town we would leave as much of the surface as possible covered by vegetation. This is in stark contrast to the ever denser land utilization now prevalent in urban areas. While it is true that there is usually some park land set aside, present planners assume that it is sufficient to have it as grassland that may double in use as playgrounds. Yet grass, while better than pavement, often dries out in summer and does not mitigate the heat island effect nearly as much as trees and shrubs. Trees have well nigh disappeared from the inner cities. Even in the suburbs or the so-called planned towns the bulldozer is the first piece of equipment to appear and in order to ease construction old trees are eradicated. Ironically, in those same areas the new houseowners plant measly little saplings in the hope that they might grow back into shade trees perhaps by the time the mortgage is paid off. Old shade trees along avenues laid out in an earlier age often succumb to diseases, air pollution, and the devouring needs of traffic.

We should reduce the needs for surface space, much of which is now occupied by parking lots. In the first place these needs will be lowered by reduced use of cars as a mode of transportation, and whatever cars remain necessary will be parked underground or under buildings. Business and apartment buildings in Metutopia would be

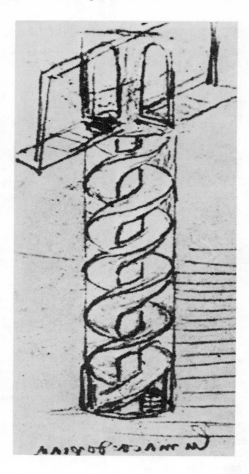

FIGURE 2. Leonardo da Vinci's "Città ideale". Communication between different levels by helix-shaped spiral staircases for pedestrians. Note the artist's propensity to write important notes in mirror writing.

required to provide for necessary parking below their establishment. Similarly, tall buildings using less land surface than low structures will be the rule rather than the exception. Pedestrian traffic will move under colonnades. Buildings will have open vegetated spaces between them. Instead of having four times the ground area of a building covered by parking space and access roads, as is often the case at present, part of the grounds will be developed as a park.

Aside from reducing the heat-island effects, the spacing of buildings will have beneficial results for ventilation. In higher altitudes during the cold season, when spare heat might do some good, much better use will be made than at present of rejected heat from furnaces, factory smokestacks, and cooling towers. Instead of dissipating this heat into the atmosphere — and perhaps producing unwanted weather modifications elsewhere downwind from the urban scene — this heat is to be channeled through smoke sewers under uncovered sidewalks, streets, highways, and bridges to melt snows that cause accidents and traffic jams. This will eliminate the need for expensive snow-clearing equipment, standby crews, and ecology-damaging salts. Centralized collection of smoke will also make deactivation of obnoxious effluents easier. In summer the

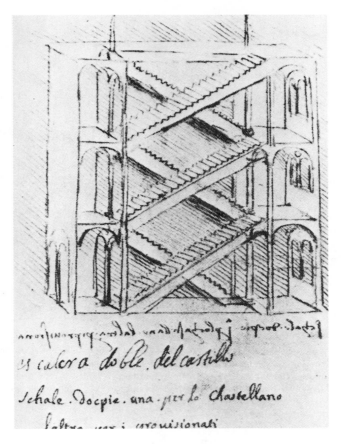

FIGURE 3. Leonardo da Vinci's "Città ideale". Communication between different levels by double staircases for porters and handymen. Note Leonardo's mirror writing.

FIGURE 4. Leonardo da Vinci's "Città ideale". Building facilities for patricians, owners of horses, and carriages. Note the mirror writing.

FIGURE 5.   Leonardo da Vinci's "Città ideale". Palaces for noblemen show the design of a penthouse with a pergola. Underneath is a mansard story for the domestic servants, followed by the dwelling unit for the noble residents. The ground level is built for receptions with access to the gardens. Menial work and cooking is carried out in the large vaults to the right, whereas the left orchard space is reserved for cellars and disposal of household waste, dog dens, or horse stables. Note the mirror writing.

heat rejection from furnaces will not be a factor, and that from other sources (such as power plants) can be diverted to horticulture, agriculture, and open-air swimming pools.

Airflow through an urban area is very complex: channeling by streets, eddies induced by buildings, and the generally much higher roughness than in rural areas cause an overall reduction in wind speeds. On an average, this is at least 10% of the speed at an airport exposure. In weak winds of less than 4/m sec^{-1}, the reduction can be as much as 40%. Obviously, these weak winds accompany synoptic stagnation situations when air pollutants accumulate and when any reduction in wind speed aggravates the problem. In Metutopia there will be no narrow thoroughfares, and distances between buildings and structures will be such as to minimize solid walls of obstacles. Uniform heights, which are common in some older parts of cities and which in essence create a new surface for the wind to move over (rooftop jet), will be avoided.

When it comes to air pollution, levels of one order of magnitude larger are almost insignificant, because values two or even three orders of magnitude higher than in rural areas often exist. In many U.S. cities there is about equipartition — if not in individual pollutants, then in the aggregate between stationary and mobile sources. The latter, of course, are primarily our cars. Stationary sources are much more readily controllable than moving ones. The construction of smoke sewers for effluents of stationary origin should be encouraged.

When it comes to cars, trucks, and buses, the heavy hand of government is trying to regulate emissions and enforce its regulations. In Metutopia the harassed commuter, the shopper, and the schoolchildren will be offered adequate, that is, frequent and cheap, high speed, nonpolluting electric transportation. The technology exists and might as well be used. The cost to the taxpayer can at least be partially recovered by reduced health costs, reduced costs of cars used for long distance travel, lower accident rates, and less time lost. Long distance traffic will be routed to and through peripheral turnpikes. These will be carefully shielded from the inhabited areas by wide belts of woodland. They serve simultaneously as recreational areas and as filters for pollutants and noise. In some regions they can also incorporate hedge structures to act as permanent snow fences. Vegetated noise barriers might also be raised in other parts of the town where such protection is needed, although the all-electric transportation system should minimize such requirements.

Noise and foreign aerosols take a high toll from the population. The former can be blamed for increases in all kinds of ailments ranging from hearing impairments to insomnia, high blood pressure, and neuropathic anomalies. The aerosols, comprising a wide assortment of substances, can be blamed for increases in the incidence of a wide assortment of diseases: bronchitis, emphysema, and lung cancer among them. They can aggravate cardiopulmonary difficulties. They may pave the way for acute upper respiratory infections, and even contribute to lead poisoning of small children. In Metutopia, aside from electrification of transportation, pollutants of stationary origin are to the maximum feasible extent controlled at the source. If this is impracticable they are ducted out of town where they are detoxified and dispersed under strict meteorological control. The role of the meteorologist as a forecaster and controller of urban pollution will in Metutopia be a central one, just as it should be now in our urban centers rather than the subsidiary position which he occupies presently. In particular he should make the decisions when fuel types ought to be switched from coal and oil to gas in certain installations, such as power plants, and when certain effluents should be stored. Such drastic measures are called for in synoptic stagnation situations, many of which can be forecast 2 or 3 days in advance.

Obviously, Metutopia will not have incinerators, but burnable trash that cannot be recycled will be burned for power generation with the meteorological safeguards indicated before. However, by and large one will look toward atomic energy for the generation of electricity, with the rejected heat beneficially used in agri- and aquaculture and effluents reduced to a minimum. Housing and buildings, according to the climatic zone in which the town is located, will be constructed to minimize consumption of energy and heat rejection. Superior insulation will be required to reduce demands for heating and cooling. In sunny areas and during summer, reflecting outside paints will increase albedos, and all glass surfaces, including windows with exposure to the sun, will have outside reflecting shutters or shading devices to avoid the trapping of heat. In cold climates and in winter, maximal heat absorption from radiation through low albedos and glass sides will be attempted. In the development of a new style of housing adapted to climatological conditions, opportunities for wholly or partially using solar energy for space and water heating will have to be exploited. Cooling problems are more difficult to handle, but in some climatic regions more underground construction is definitely to be encouraged.

II. ECONOMIC PROBLEMS

A. Power Supply

1. Power Generation

The demand for services provided by public utilities, such as fuel oil, electric power, radio, television, telephone, water, and gas, is surely affected by variations in the weather and the climate. The normal seasonal climatic variations create associated seasonal variations in the supply of water for hydroelectric generation plants, and similar variations in the demand for fuel oil, gas, and electricity. Possibly of even greater economic significance are the day-by-day variations in the weather. Air conditioners can create a critical demand for the supply of electric power in some hot and humid urban areas, particularly in the south and east of the U.S. The maintenance of services in adverse weather conditions is also a factor of some concern to utility companies, particularly since the demand for some services increases considerably in critical weather conditions. This concerns especially power waste created by lightning.

2. Fuel Consumption

The heating required to maintain a comfortable indoor temperature depends on controllable factors, such as the design and size of the building, and also the "uncontrollable" factors of the external climate, the most important of which are temperature, wind, and humidity. For example, if the outside air is cold, the wall temperatures of a building will be low, and thus it is necessary to keep the indoor air temperature a few degrees higher than "normal" in order to compensate for the radiative loss of heat. Additional fuel is consumed because of this loss of heat. In fact, the difference between the indoor temperature and that of the outside air is clearly associated with the amount of heating required; the concept of the heating degree may therefore be particularly useful in heating requirement calculations. Conversely, in summer, if the outside temperatures are high, it is desirable to cool buildings because they acquire part of the outside heat.

Other weather features influencing the heating requirements of buildings and hence fuel consumption are evaporation and wind, the latter being particularly important in areas subject to severe winters. Moreover, the actual siting of a building has an appreciable effect on heating requirements, which for buildings outside a built-up area, are estimated to be 15% to 25% higher than for similar structures within the built-up area.

3. Organizing Electric Supply

Weather conditions have an important influence on the generation, transmission, and distribution of electric power. The weather affects the load on the plant and in addition, it may cause damage to equipment and interrupt the supply of electricity to the consumer. The heating and lighting components of the total load are also closely associated with weather conditions, since the lighting load is a function of the daylight illumination and, apart from "abnormal" weather conditions, it varies throughout the year in a predictable manner. The heating load on the other hand varies with the air temperature, and to a lesser extent with the wind, sunshine, and humidity, and although it, too, changes in a predictable manner throughout the year, it can be irregular as a result of sudden cold spells.

Careful study of the climatic conditions affecting air electricity is vital to the choice of location of a hydroelectric undertaking. In particular, attention should be given to precipitation, including the incidence of snowfall and the duration of snow cover, the frequency of droughts, and the behavior of streams within the catchment. Second, the frequency of strong winds, humidity, and ice conditions need to be considered in the design of high-voltage line structure, and maximum temperatures are taken into account when the degree of clearance of line-sag during summer months is forecast. Moreover, an electricity supply system is required to respond instantly and unfailingly to the ever-changing demands of its consumers, and weather conditions have a very large bearing on this demand.

The importance of short-range forecasts should not be overlooked. For example, at around freezing point the increase in demand per 1°C sustained fall of temperature can be as much as 160 MW, and at still lower temperatures this can rise to 200 MW with no signs of demand saturation. Other meteorological elements, such as wind, cloud, fog, and precipitation, also cause considerable variations in electricity demand.

Associated with this is the importance of distinguishing the effect of weather and climate factors on power consumption at different times during the day. All of them depend intimately on the adverse effects of the ambient temperature.

4. Gas Consumption

In many parts of the world heating requirements of houses, factories, and office buildings are furnished by gas rather than electricity or oil. The natural gas industry of the U.S., for example, is not only an important factor of the utilities division of the economy, but it is also particularly weather-sensitive.

As with electric power, the value of weather lies principally in its connection with consumer demand, but in the case of the natural gas industries it is particularly important to be able to anticipate in advance what the consumer demand will be. The delivery of natural gas over long distances cannot ordinarily be increased as rapidly as can electrical supply. It is therefore very important to be able to anticipate future needs in the light of weather forecasts, since the demand for gas resulting from colder temperatures must be met by dispatching it from a remote source before the low temperatures occur. Consequently, gas supply companies constantly refer to the meteorological conditions and forecasts for the area being served, and several gas companies have found it profitable to employ meteorologists who can interpret weather conditions specifically in terms of the problems of gas distribution.

B. Housing
1. Architectural Design

Amongst a number of studies on the relationship between climate and architectural design, the contribution by Aronin[11] is particularly noteworthy for specific climate-design applications. In addition, a "Bibliography of Weather and Architecture"[14] should be noted. In general, however, buildings and structures are usually designed to withstand the probable combinations of climatic extremes, and to make indoor conditions comfortable and healthy regardless of the weather conditions outside.

The relevant weather factors in architectural design are thunderstorms, temperature, sunshine, humidity, wind, and precipitation, the selection of the design data usually being made in accordance with the probability of the occurrence of the most severe weather conditions that will justify a specific design. The probability value to be used depends upon a number of factors, including the degree of climatic control required, how much the structure itself is expected to control, and the use that is to be made of mechanical aids such as heating or cooling equipment. In addition, account must be taken of both the daily and the seasonal variations in sunlight and any drainage or roof stress problems that could arise under extreme rainfall or snowfall conditions.

2. Glass Buildings

In cold climates considerable interest is shown in the use of glass in buildings, and this has generated a considerable difference of opinion between architects. The ideal arrangement for solar control is to design the building with the glass facing south with the provision of both a canopy shade and a venetian blind. The purpose in the design of the "solar house" is to attain more efficient use of the sun for light and heat. The effect of such a house on lighting is self-evident. The heat can, however, be utilized in two different ways: first, by trapping the sun inside the house, using large-paned windows of two or more glass layers; and second, by using the sun's heat to warm a liquid or solid solution, which is then employed as a fuel to heat the house. This latter effect is particularly significant from an economic viewpoint, and reports from the owners of "solar houses" suggest substantial fuel savings over conventional methods. Thus, although solar radiation, particularly as a means of heating, has been largely ignored in building design, there can be no doubt of its increasing significance in the future.

The adverse aspect of a glass building is its lack of ions and electric fields which has been termed a "death trap".[12]

3. Economic Impact

The impact of weather on the construction industry in the U.S. has been reviewed in a number of recent studies, the interest in such studies being justified since weather produces a severe operational and economic impact on the construction industry. This is caused by three factors, namely: (1) a large percentage of the construction operations is sensitive to adverse weather conditions; (2) building construction is often vulnerable to ordinary weather occurrences such as winds, snow, thunderstorms, etc., as well as to some extraordinary ones such as hurricanes, tornadoes, etc.; and (3) building materials and systems are susceptible to a greater or lesser extent to the degradating effects of the natural weathering processes.

The advent of modern plastic designs in housing has created new aspects with regard to static electricity. The latter can severely affect the well-being of the inhabitants of the modern house, and on the other hand, may have an adverse effect on the wear and tear of the building materials.

C. Transportation

1. General

Almost all segments of transportation (air, water, land, and pipeline) are affected by the atmospheric and electric environment. The day-to-day weather conditions, for example, affect ship, bus, railway, and airline schedules, and adverse weather conditions increase the number of accidents in most kinds of transportation. The climate is also important for, and climatological data can assist, planners associated with design problems in transportation facilities and transportation equipment, from automobiles and highways to spacecraft and their associated ground facilities.

There are generally three kinds of weather effects associated with transportation. First is the influence of the actual weather conditions encountered, such as the weather on a flight path from one place to another. Second is the use, particularly in planning, of weather forecasts; and third is the importance of weather modification, the clearing by man of fog at airports being a notable example. As will be appreciated, the airline industry is perhaps most affected and influenced by all three aspects of the weather, but other segments of the transportation industry, including long-distance road transport, city taxi services, city bus services, transportation by pipeline, and the routing of ships, are all weather-sensitive.

2. Air Transport

The weather phenomena that most frequently influence airline operations are those involving visibility and runway conditions. In addition, there is the effect that tornadoes, thunderstorms, and other severe storms have on aircraft, but this impact can usually be considered small due to the short duration and small areal coverage involved. Moreover, present-day communications, weather radar, and improved forecast ability usually allow airlines to cope with these conditions, and the costs of the necessary equipment are not excessive when the savings are considered. The frequency of fog, snow, blizzards, slush, water, loose snow, and ice on runways has, however, a far greater effect on operations, this being partly due to their more common occurrence and their general areal extent. In addition, a more severe and increasingly frequent weather event for high-altitude aircraft is clear-air turbulence; a considerable amount of damage in some aircraft is attributed to such turbulence. Another effect of weather on airline operations is the surface temperature and static electricity on runways, affecting takeoffs and at times restricting loads and hence airline revenue.

Most aspects of weather are fortunately only an economic hazard to airline companies, and the direct physical destruction by weather of an aircraft en route or on landing is rare. However, there have been many airline disasters, some of which are believed to be directly associated with adverse weather conditions. It should nevertheless be appreciated that crashes are only a few among the millions of flights safely completed each year all over the world. Moreover, as better methods are found to give an even more accurate picture of the weather, pilots will be able to lessen the chances of weather- and air-electricity-associated air crashes. This could avoid mishaps like those encountered by the abortive rescue mission of the U.S. Air Force in Iran in 1980.

Of the various weather elements that affect air transportation, none has such a marked effect as fog. The losses and inconvenience caused by fog at airports can usually be reduced by three methods: first, provision of electronic landing aids which leaad to safe "zero-zero" landings; second, accurate terminal weather forecasting; and third, clearance of fog through weather modification. All these methods have improved considerably during the 1960s, but there are still limitations with each method.

3. Water Transport

For centuries man has been aware of the effect of the weather upon water transport, and although modern navigational aids and improved safety procedures have considerably reduced marine disasters, it is still not uncommon for ships to be lost at sea with little if any trace. Moreover, the power of the sea is such that in certain circumstances man unfortunately can only stand and watch.

In earlier times, sailing ships were very dependent on the prevailing wind conditions. Today, the importance to shipping of the prevailing winds is minimal, and although strong winds associated with tropical revolving storms and other severe cyclonic storms can at times be destructive, the main weather elements affecting water transport are low temperatures and fog.

Low temperatures influence several operations including loading and unloading activities, extra protection for both cargo and passengers often being required. In addition, if ice forms, navigable waterways may become impassable.

Fog is another important factor affecting water transport and although lights, foghorns, radio, radar, and electronic devices for sounding ocean depths have considerably lessened the danger of collision or running aground, dense fogs still bring harbor traffic to a standstill and contribute to many marine accidents. Nowadays, the oil tankers present a special hazard for water traffic.

Precipitation is not a major factor affecting shipping, but may disrupt loading and unloading operations of perishable cargo. At some harbors special loading equipment is used to counteract wet weather conditions.

A natural outcome of man's response to weather and sea transportation has been the provision of specialized weather forecasts which enable a ship to gain all the advantages of the prevailing weather conditions, and to avoid most of the disadvantages. The technique is known as "weather routing" which according to Evans[13] means to project the wave pattern ahead as far as possible along the normal route of the vessel from the point of departure to that of destination, utilizing the 3- to 5-day forecast charts of surface weather and wave height, modified perhaps by local prognosis. Estimates are then made of the vessel's speed as a consequence of encountering predicted wave heights based upon previously determined performance data of the ship under similar conditions. By repeating this process and projecting ahead, the ship's least-time track is computed, which may be adjusted from time to time.

The use by ships of weather information provided by weather consultants, rather than using the weather information provided by the various government weather services, also shows the value that shipping companies place on accurate weather information. Special studies have shown that the threat from heavy weather damage has been reduced by 46% for vessels utilizing recommended routes. Other benefits also accrue, such as fuel savings and ultimately insurance costs.

Air electricity, especially thunderstorms with their lightning, can also severely affect shipping. Tankers with inflammable cargo are especially prone to the devastating effects of thunderstorms.

4. Rail Transport

In some areas of the world, snowdrifts have caused passenger trains to be stranded for days, and few railway lines in the world have not at some time in the past suffered from the vagaries of the weather. Thus, although railways would seem to be little affected by weather because of their permanent all-weather track systems, they are far from immune to weather hazards, and their operations must be constantly geared to weather changes. Severe storms, for example, may damage tracks, bridges, signals, and communication lines; floods, heavy snowfall, and earth slides are related menaces, and heavy snow and avalanches are particularly troublesome in the mountains. In some areas, for example, it has been expedient to build expensive tunnels to avoid the severe weather of high altitudes. Low visibility due to fog or precipitation are additional weather factors which call for extra caution and decreased speeds, and some passenger and freight schedules can be disrupted by the effects of such weather on tracks and equipment. Weather may of course cause dislocation at any time of the year, and in extreme cases bridges can be washed away and major subsidence and landslides may occur, the resultant diversions and rebuilding adding greatly to the running costs of railways.

Weather conditions which deviate from the optimum point can in fact affect railway systems in two major ways. First are those conditions which hinder the safe and rapid movement of the trains, and probably the most obvious and most costly weather factors detrimental to railway operations are those due to low temperatures, which cause problems of track expansion and contraction. For example, long lengths of rail not properly fastened down can easily loosen or buckle as their length decreases or increases, and in the case of manually operated switches and semaphore signals, lightning can cause settings to change and as a result may lead to accidents.

A set of more serious problems can occur with ice, for any water left on the railbed as a result of poor drainage will freeze where it lies and this may in turn cause operational problems with the switches and make the tracks heave up with its expansion. To counteract the freezing in switches, antifreezing compounds can be spread on the slides along which the switching rails move, or in some cases expensive heaters can be put in operation. Freezing may also directly affect the train itself, and high-powered heat sheds are now being used in some areas for loaded mineral carriers to counteract ice-bonding.

Apart from the effects of weather which may determine the actual safe movement of the train, there are various economic factors which are associated with freight and passengers. For example, to maintain passenger cars at a comfortable temperature, expensive heating and air conditioning apparatus must often be installed. Similarly, freight, like passengers, has to be protected from the weather, but to differing levels. Perishable goods, for example, must be kept at specific temperatures; accordingly, railways have invested heavily in insulated heated or refrigerated freight cars. Humidity, too, must be kept at a specified level for both food products and for metals which could easily be corroded by condensation of water vapor, while the possibility of expansion of liquid or gaseous freight must also be taken into account with some payload space being reserved for possible freight expansion.

Railway administrators are not yet sufficiently familiar with the possible effects of air electricity on plastic devices and coatings in Pullman cars. This is a topic which may need more elaborate research.

5. Road Transport

A major consideration in the relationship of weather to road transport is safety, since poor visibility, slippery surfaces, and gusty winds greatly increase the hazards of driving, a factor which is clearly borne out by accident statistics. Increased traffic in the public sector also occurs during times of adverse climatic conditions, particularly bus and taxi services in cities. Although a heavy shower in the midafternoon creates a heavy demand for cabs and buses, fog produces so many headaches and possibilities of wrong turns for the cabby that profit is negligible.

It is generally recognized that the factors tending to reduce traffic include rain, low temperatures, snow and fog, while sunshine and high temperatures usually tend to increase it. Temperature is more important in winter than in summer, and sunshine and rainfall are more important in summer than in winter.

Other surprising facts are (1) inter-urban traffic appears to be more sensitive to weather than does urban traffic; (2) warmer and sunnier than normal conditions appear to be associated with increased traffic flow, whereas colder and wetter conditions than normal appear to be related to decreased traffic flow; and (3) traffic flow appears to be most sensitive to the weather at the weekends.

Another relationship between weather and highways is that of the many maintenance problems and flood damage caused by various elements of the weather. A contributory factor to impeding traffic is snow, and although engineers have found it practical and economical to melt snow as it falls, using special heaters under footpaths, driveways, runways and ramps, snow still presents a formidable problem to many municipalities and traffic control authorities. Snow tires, tire chains, shovels, snow brooms, scrapers, sand, salt, and numerous ice-melting compounds are expensive. However, the snow menace can bring to a standstill many U.S. and other world cities.

The effect of weather on the pedestrian is also important, major weather factors being precipitation, fog, and sunshine. For example, the glare from dazzling white pavements can be most pronounced, and Griffiths[14] suggests that the use of colored pavements, such as green or pink, would reduce this annoyance, and at the same time lead to an increased absorption of sunshine which would be especially useful in winter for melting snow. An associated factor is the protection of pedestrians in shopping areas from rainfall or sunshine.

The automobile in its use and operation is subjected to many weather conditions, especially sun and heat. Such conditions have a direct effect upon the automobile's design and construction, and an engineer therefore cannot ignore the climatic factor when designing an automobile because nearly all its functional parts are influenced by the ever-changing climatic elements. Among the major weather elements is wind, which constitutes a most significant effect upon the body design and steering geometry of a vehicle. Therefore, engineers strive towards a design that when encountering a wind will not make the car deviate perceptibly within the reaction of the driver. The production cost of automobiles is also affected by weather conditions, many of which are now taken very much for granted. The use of ionizing apparatus to improve alertness of weather-prone drivers is a welcome improvement.

D. Manufacture

1. Industry

In comparison with agriculture the number of studies completed on the effect of weather and climate on manufacturing is minimal. The diverse impact of weather and climate on industry may be categorized into two broad divisions: those influencing the location of industry and those affecting the operations of the industrial plant once an industry has been established. Associated with these factors are the design of the plant, and the planning of its operations.

2. Operational Problems

Weather conditions are also reflected in almost every phase of manufacturing; moreover, the impact of various weather conditions, such as storms, can cause workers to arrive late for work, hamper essential outdoor activities, damage perishable goods and equipment, or interrupt power. Wind, air electricity, and relative humidity are also sometimes important in that these conditions may increase the fire hazard at manufacturing plants, especially if combustible or explosive materials are being processed or stored.

Textile factories need negative ionizing rods to dispel the positive air charge accumulating on the rayons which otherwise would severely hamper the process of interlacing.

3. Supply Problems

Initially, the importance of climatic factors rather than weather problems faces industry. For example, such factors as the availability and flow of water, the seasonal conditions affecting heating, cooling and dehumidifying of the air, and the atmospheric stability or instability of the site in relation to air pollution are directly related to capital investments. Labor, fuel supply, raw materials, markets, land cost, transportation facilities, and local or state taxes must also be taken into account, in addition to their direct association with climatic conditions. Climatological components which influence storage, warehousing, and the operation of transportation facilities are also important. Furthermore, almost all industries with outdoor activities are to a greater or lesser degree subjected to climatic variations, transportation, construction, aircraft manufacture, ship building, and strip mining industries often being affected by freezing, fog, snow, ice, lightning, and gales.

E. Commerce
1. Business

Although a glance at most newspapers will indicate the importance of weather conditions to the world of commerce, few if any substantial studies have been completed on the specific effects of weather on the business community. Indeed it seems almost paradoxical that the value of the weather to business and commercial activities seems to be ignored not only by the meteorologist and climatologist, but also, rather surprisingly, by the businessmen themselves.

Many economic, social, psychological, and environmental factors influence the state of economic activity at any given place in time, and of the environmental factors to be considered, those concerned with the atmospheric environment are of paramount importance. In spring the sale of gardening supplies and outdoor sports equipment increases sharply during the first few warm days. The housewife buys more salad vegetables and "lighter" items for family meals. On a sunny day in summer, the sales of soft drinks, ice cream, cold meats, and other picnic items are likely to be heavy. With the first cool spells of fall, sales of fuels begin to rise and the demand for furnace repairs or installation grows. The first threat of freezing weather is accompanied by a rush of motorists to garages and service stations for antifreeze additive. The effects of cold weather and storms on business extend into the winter. Fuel consumption rises, and after a hard freeze, the plumber makes his rounds to thaw out pipes.

Air ionization also has an adverse impact on business. In countries suffering from ionizing hot winds everybody tries to postpone business on days with such winds.

2. Banking

The recognition of the seasonal nature of economic activity in any area is of prime interest to the banking community. Commercial banks, for example, are often faced with seasonal demands for credit, and these demands do not always coincide with seasonal variations in new deposits. In particular, banks often find that the demands for loans vary with changes in business conditions in their communities, and as has been suggested, at least part of these variations in business activities are related to seasonable weather conditions. Indeed, it would appear that dividends are almost there just for the asking, provided the world of business and the world of meteorology would only get together and really begin to appreciate how and in what specific way the atmospheric environment influences our activities and mode of living.

A typical observation is the unrest and irritability at the Stock Exchange on days when positive ionization runs high and people are upset by serotonin release. Many investors have lost their fortunes on such days.

3. Shopping Centers

A step in overcoming weather conditions is the shopping center in which a major department store is connected to smaller shops by an enclosed "air-conditioned" mall so that the shopper is not only protected from precipitation but also from heat or cold. Weather might affect the sales of a retail store in four ways: first, the weather could be of such a nature that it is for one reason or another "uncomfortable" to go shopping; second, the weather could produce situations that would physically prevent people from going to the store, as in the case of snowdrifts over roads and streets; third, the weather may have psychological effects on people that may change their shopping habits; and fourth, some kinds of merchandise may be more desirable during a period in which certain types of weather prevail.

The commodities market is the exchange where commodities such as wheat, corn, soybeans, meat, eggs, potatoes, sugar, cocoa, and frozen fruit juice are bought and sold; not directly, it should be added, but rather the rights to buy and to trade (and hence sell) a certain quantity of these goods for delivery at a stated time in the future. The prices for the commodity "futures" are related to a number of factors, but possibly of greatest importance is the anticipated demand for and the anticipated supply of the particular commodity at some future date. Consequently, it can be seen that there is at times a very close relationship between the weather conditions and the price at which "futures" in a commodity will be sold. Thus for instance, people would refrain from buying perishable commodities on days when hot weather may spoil the quality of the produce. The best method of cooling cannot always prevent bacterial multiplication of dairy products on hot days.

4. Weather Insurance

The weather is a very important factor in many areas of the insurance industry. In the U.S., property valued at more than $500 billion is insured against storm damage (that is almost $2500 per person), and the annual paid claims range from $0.5 to $1.0 billion. With this much money involved it is evident that weather and insurance are closely associated.

Many kinds of weather-related insurance are issued each year, but possibly of most direct concern to the policyholder is crop insurance, since crop failure through adverse weather conditions is an ever-present risk in most kinds of farming in many areas of the world. Fortunately, because of improvements in technology, the risks of crop failure attributable to natural hazards, such as thunderstorm, drought, disease, or insects, are smaller now than a decade ago, but the fact remains that when failure does occur today it is usually much greater because of the increase in costs per acre of producing crops. Many natural risks are of course insurable, and two general types of crop protection are often available to farmers: the first is crop-hail insurance, which is offered in the U.S. by both stock and mutual companies, and the second is all-risk crop insurance, which is offered in the U.S. by the Federal Crop Insurance Corporation (FCIC), an agency of the U.S. Department of Agriculture.

REFERENCES

1. **Watson, D. J.**, Weather and plant yield, in *Environmental Control of Plant Growth*, Evans, L. T., Ed., Academic Press, New York, 1963, 337.
2. **Thompson, L. M.**, Weather and our food supply, *Cent. Agric. Econ. Devel.*, Iowa Sate University, Ames, Rep. 20, 1964, 1.
3. **Krueger, A. P.**, The biological effects of gaseous ions, in *Aeroionotherapy*, Gaultierotti, R., Ed., Carlo Erba Foundation, Milan, 1968, 65.
4. **Lemstroem, S.**, Electricity in agriculture and horticulture, *The Electricians*, London, 1904.
5. **Sidaway, G. H.**, Some early experiments in electriculture, *J. Electrostat.*, 1, 389, 1975.
6. **Sulman, F. G., (in collaboration with) Ben-David, M., Danon, A., Dikstein, S., Givant, Y., Khazen, K., Mishkinsky-Shani, J., Nir, I., and Weller, C. P.**, *Hypothalamic Control of Lactation*, Monogr., Springer-Verlag, New York, 1970.
7. **Landsberg, H.**, Inadvertent atmospheric modification through urbanization, in *Weather and Climate Modification*, Hess, W. N., John Wiley & Sons, New York, 1974, 726.
7a. **Landsberg, H.**, Drought, a recurrent element of climate, *W.M.O. Spec. Environ. Rep.*, No. 5, 45, 1975.
8. **Winkle, S.**, Leonardo's Citta ideale, *Muench. Med. Wschr.*, 117, 99, 1975.
9. **Landsberg, H. E.**, Climates and urban planning, *W.M.O. Tech. Note*, No. 108, 364, 1970.
10. **Landsberg, H. E.**, The meteorologically utopian city, *Bull. Am. Meteorol. Soc.*, 54, 86, 1973.
11. **Aronin, J. E.**, *Climate and Architecture*, Van Nostrand Reinhold, New York, 1953.
12. **Griffiths, J. F.**, *Applied Climatology*, Oxford University Press, London, 1966, 1.
13. **Evans, S. H.**, Weather routing of ships, *Weather*, 23, 2, 1968.
14. **Griffiths, J. F. and Griffiths, M. J.**, A Bibliography of Weather and Architecture, ESSA Tech. Mem. EDSTM No. 9, U.S. Department of Commerce, Silver Spring, Md., 1969, 1.

Chapter 3

LONG-TERM CLIMATE CHANGES

I. SYNOPSIS OF PALEOCLIMATOLOGY

A. Early Earth Ages

The understanding of paleogeology and paleoclimatology is only possible with the aid of an elaborate synopsis which is given in Tables 1 and 2. The following text should facilitate reading of the details given in the tables.

The entire span of Earth history is estimated at more than 3000 million years. Probably four fifths of it had elapsed before the first appearance of primitive flora and fauna in the Cambrian epoch.

1. Archeozoic Period

Earth history begins with the Archeozoic period 1000 to 3000 million years B.P.* The absence of any life sign has given the earliest part of this period the name of Azoic epoch. It was characterized by ancient crystal lime rocks, mountain building, and vulcanism (many events obscured by vast lapse of time), and primitive forms of marine life, perhaps algae-like, but with no direct fossil evidence.

2. Proterozoic Period

The Proterozoic period followed the dawn of Earth, first with the Algonkian and later with the Precambrian epoch, lasting from 600 to 1000 million years B.P. It was characterized by much mountain building, metamorphism of rocks, vulcanism, and primitive marine life, mainly without shells, leaving only meager fossil remains.

3. Paleozoic Period

The Paleozoic period together with the two preceding periods were once conveniently called Primary era — a name now rarely used. The Paleozoic period comprises five epochs in succession: Cambrian, Ordovician, Silurian, Devonian, and Carboniferous ("The Coal Age").

The Cambrian epoch — This was characterized by widespread submergence and deposition of sedimentary rocks and the first appearance of abundant fossils, mainly of shelled marine invertebrates (mollusks and trilobites). It lasted from 500 to 600 million years B.P.

The Ordovician epoch — This was characterized by sediments deposited, mountain building in New England and Canada, and abundant mollusks and trilobites, early forms of fish, and no evidence of land animals or plants. It lasted from 435 to 500 million years B.P.

The Silurian epoch — This was characterized by widespread development of plains by erosion and by emergence, development of fishes, first land animals (spider-like), first land plants, and abundant corals. It lasted from 395 to 435 million years B.P.

The Devonian epoch — This was characterized by widespread submergence — mountain uplift and vulcanism in New England, abundant fishes with vertebrae and paired fins, the first amphibians, and the first forests (tree ferns). It lasted from 345 to 395 million years B.P.

The Carboniferous epoch — This contained the Mississipian and the Pennsylvanian ages, the earlier one being characterized by widespread submergence and deposition

---

* B.P. = before present: 1952.

Table 1

GEOLOGICAL ERAS AND THEIR IMPACT ON CLIMATE, FAUNA, FLORA, AND MAN

Era	Period	Epoch	Typical features	Estimated age B.P. thousand years ago
Quaternary era of Man and recent life	Holocene	Iron Age	Preparation of iron tools	1.2
		Little Ice Age	Horses, elephants, pigs	1.5—1.85
		Chalcolithicum = Bronze Age	Use of copper alloys	2—3
		Neolithicum	Polished stone implements	3.5—5
		Mesolithicum	Primitive flint implements	10
	Pleistocene	Alluvium	*Homo sapiens* and recent fauna	40
		Diluvium	Solo Man, mammoths, bovids, big birds, marsupialia	100
	Upper Pleistocene	Ice Ages	Neanderthal Man, boreal fauna	110
		Wuerm Glaciation	Rhodesian Man using blades	120
		Neo-Paleolithicum	Cro-Magnon Man using punch stones	150
		Interglacial Age	Achuleon Man using hammer tools	200
		Meso-Paleolithicum	Fontochevade Man using flake tools	240
	Middle Pleistocene	Riss Glaciation	Arago Man using hand axes	300
			Swanscombe Man using stone axes	350
		Interglacial Age	Steinheim Man using fire	400
			Vertesszoeloes Man using pebble tools	440
		Mindel Glaciation	Peking Man, *Homo erectus* Sinanthropus, rich mammalian fauna	480
		Interglacial Age	Heidelberg Man using pebble tools, hunting, and using fire	550
		Guenz Glaciation	Java Man, *Pithecantropus erectus* developing primitive horticulture	600

Table 2
GEOLOGICAL ERAS AND THEIR IMPACT ON CLIMATE, FAUNA, FLORA, AND MAN

Note: Details on the different ages are given in the geological glossary.

Era	Period	Epoch		Typical features	Estimated age B.P. million years ago
		Lower Pleistocene	Interglacial Age	*Homo habilis* using cloths, Olduval and Lake Rudolf Man	800
			Donau Glaciation	Transformation of homines to erect primates (Sterk Fontain Man)	1000
Tertiary era of mammals and "modern" life	Cenozoic	Pliocene		Transition to ice epochs, large carnivorous animals	1—10
		Miocene		Moderate climate, whales, apes, and grazing animals	10—25
		Oligocene		Warm climate, first monkeys and apes	25—40
		Eocene		Warm climate, horses, elephants, modern flora	40—60
		Paleocene		Warm climate, development of modern fauna	60—63
"Secondary" era of continental drift and "medieval" life	Mesozoic	Cretaceous		Cold climate exterminates dinosaurs, ammonites, modern flora begins	63—135
		Jurassic		Dinosaurs, crocodiles, flying reptiles, first birds, mammals, insects, conifers, ferns	135—180
		Triassic		Dinosaurs, Ichthyosaurs, reptiles, marine animals, coniferous forests, evaporite rock salts	180—225
		Permian		Glacial climate first, moderate later, reptiles, insects, conifers	225—280

Table 2 (continued)

GEOLOGICAL ERAS AND THEIR IMPACT ON CLIMATE, FAUNA, FLORA, AND MAN

Era	Period	Epoch	Typical features	Estimated age B.P. million years ago
"Primary" era of "ancient" life	Paleozoic	Carboniferous	Warm climate first, glacial later, sharks, winged insects, spiders, conifers, ferns, club moss	280—345
		Devonian	Moderate climate first, warm later, amphibians, fishes, insects, forests	345—395
		Silurian	Warm climate, corals, first land plants and animals	395—435
		Ordovician	Moderate-warm climate, no land animals, first corals and fishes	435—500
	Proterozoic	Cambrian	Cold climate first, warm later, first star fish and foraminifers, snails, shellfish, mollusks	500—600
		Precambrian Algonkian	Glacials covering whole Earth, first algae, radiolaria worms	600—1000
	Archeozoic	Azoic	Before the dawn of life	1000—3000

Note: Details on the different ages are given in the geological glossary.

of sediments and development of sharks and other fish, numerous amphibians, abundant forests of ferns and primitive conifers; the later one by fluctuating seas in the interior, formation of extensive swamps, vast forests of fast-growing trees and other plants, and complex marine life with the rise of reptiles and insects. Both together lasted from 280 to 345 million years B.P.

The whole Paleozoic period lasted approximately 320 million years, i.e., about 20% of all geologic time.

4. Mesozoic Period

This period, once conveniently called the Secondary era of "medieval" life and continental drift, comprises three epochs with an earlier one, the *Permian,* serving as a transition epoch between the *Paleozoic* and the *Mesozoic* period. The Permian epoch was characterized by general emergence, folding of Appalachian mountains, widespread aridity, and decline of fern trees and rise of conifers, great variety in reptiles and insects, and the disappearance of many marine invertebrates. It lasted from 225 to 280 million years B.P.

The Triassic epoch — This was characterized by large land areas, continuation of aridity, some rocks of land-deposited origin, and vulcanism; diverse and abundant reptiles (crawling, walking, flying, swimming), many complex marine animals, and mainly coniferous forests. It lasted from 180 to 225 million years B.P.

The Jurassic epoch — This was characterized by Appalachian mountains base-leveled, Pacific coast vulcanism, and submergence from Colorado to Alaska; giant reptiles (dinosaurs), the first birds, primitive mammals, and many insects similar to present forms. It lasted from 135 to 180 million years B.P.

The Cretaceous epoch — This was characterized by the last great submergence, from the Gulf of Mexico to Alaska, which was followed by upheaval and the beginning of the Rocky Mountains, rise of mammals and birds, the decline and extinction of dinosaurs, and the development of modern flowering plants and deciduous trees. It lasted from 63 to 135 million years B.P. and received its name from the chalk and coal deposits originating from this time.

The whole Mesozoic era lasted approximately 160 million years, i.e., about 8% of all geologic time.

B. Tertiary Era

Cenozoic period is a synonym for the Tertiary era, a name popular enough to still be retained. This is the era of modern life during which Man and the mammals emerged. It is divided by some authors into the earlier *Tertiary* and the later *Quaternary* era. Both together lasted approximately 65 million years, i.e., about 4% of all geologic time.

The Tertiary/Cenozoic period began with the *Paleocene,* followed by the *Eocene, Oligocene, Miocene,* and *Pliocene.* During these five epochs the following decisive changes took place: elevation of Rocky, Sierra Nevada, and Cascade Mountains, the Colorado and Columbia plateaus, and the Great Basin; development of mammals (primitive types of elephants, horses, deer, cats, dogs, whales, and many others, including the first apes), birds, and trees similar to modern types. The Paleocene lasted from 60 to 63 million years B.P., the Eocene from 40 to 60 million years B.P., the Oligocene from 25 to 40 million years B.P., the Miocene from 10 to 25 million years B.P., and the Pliocene from 1 to 10 million years B.P.

C. Quaternary Era

The Quaternary Cenozoic period is the time of our present life. It contains five ice ages (glacials), four interglacial ages, and has many subdivisions. Counting from the earliest period, it encompassed the *Lower Pleistocene* period 800 to 1000 thousand years B.P., the *Middle Pleistocene* 240 to 800 thousand years B.P., the *Upper Pleisto- cene* 110 to 240 thousand years B.P. (also called *Diluvium*), and the *Holocene* 0 to 110 thousand years B.P. (also called *Alluvium*). During this time the following devel- opments took place: present tectonic and gradational land forms, the California Coast Mountains appeared, the Great Ice Age; the development and dominance of primitive man, new species of plants and animals, and the appearance of intelligent man. Man developed flints (*Mesolithicum*), polished stone implements (*Neolithicum*) at 3.5 to 5000 years ago, followed by the Bronze Age (*Chalcolithicum*) at 3000 years ago, and by the *Iron Age* approximately 1200 B.C. The whole Cenozoic period embraced ap- proximately 70 million years, i.e., 4% of all geologic time. During this epoch there was also a *Little Ice Age*, from 1430 to 1850 A.D. (see Chapter 4, Figure 2).

D. Darwinism

The discussion whether Man developed from apes or by a divine act of creation looks futile today. All living creatures developed from a primordial cell, branching off in thousands of species, and there is no proof that the homines developed from apes or rather from another species. The hominids which branched off directly from the primordial cell developed directly into the species of *Homo sapiens*.[1]

II. DATING PROBLEMS

A. Nomenclature

Efforts to create systems of classification based on superposition began about the middle of the 18th century. Giovanni Arduina, an Italian worker, proposed in 1760 that the rocks of the Earth be divided into Primary (first), Secondary (second), and Tertiary (third) groups. Later on, the Quaternary (fourth) was introduced as a com- panion term to the others to include the very youngest soils and alluviums that are not solidified to rock. It heralds the appearance of Man and mammals. The terms *Primary* and *Secondary* have been dropped by most geologists, but *Tertiary* and *Quaternary* are still used.

The stratigrapher has a peculiarly complex problem in attempting to set up a series of divisions and subdivisions of his material. These complexities are fully expressed in the various reports of the American Commission of Stratigraphic Nomenclature (1949 to 1961) and the International Subcommission on Stratigraphic Terminology (1952 to date). From the earliest times geologists have used separate sets of terms for the divi- sion of geological time into convenient portions and for the rocks belonging to these time divisions. A complication has been introduced by the establishment of a separate set of terms based upon the fossil assemblages of the rocks. These three categories have been referred to as chrono-stratigraphic or geological time units, time-rock units, and bio-stratigraphic units. The term litho-stratigraphic (rocks stratigraphic) has been introduced to describe units defined in terms of lithology. A substantial number of geologists regard bio-stratigraphic units and chrono-stratigraphic units as identical.

The aforementioned International Subcommission has recommended the following terms for geological time units and chrono-stratigraphic units:

Rank	Geological time units	Chrono-stratigraphic units
1st order	Era	(Erathem)[a]
2nd order	Period	System
3rd order	Epoch	Series

4th order	Age	Stage
5th order	(Time)[a]	Substage

[a] These terms are hardly ever used.

B. World Calendar Discrepancies

Astronomical measurements give the period of the passage of the Earth around the Sun as 365 days, 5 hr, 49 min, and 46 sec (365.2422 days). The modern Christian calendar, called the *Gregorian Calendar* since its enactment by a decree of Pope Gregory XIII in March 1582, adopts a 365-day year with 1 day added to the month of February every fourth year (leap year) except for the century years, which are only leap years when the number is divisible by 400. It is suggested that all millennia except those divisible by four be leap years, in which case the calendar would keep constant phase to within 1 day in 20,000 years.

Many early calendars took as a basic unit the lunar month (about 29.5 days), the recurrences of new moon and full moon being the easiest way of keeping track of the progress of the year in readily followed steps. The months were alternated between 29 and 30 days. Unfortunately, the year of the Earth's orbital passage around the sun is not perfectly divisible in this way: about 11 days have to be added if the seasons are to be kept at constant dates. The Muslim calendar ignores this and undergoes a complete cycle of phase change with respect to the seasons every 34 lunar years, which approximate 33 solar years. Thus, the Muslim calendar has ever-changing dates of feasts. The Jewish calendar catches up with the solar year every 3 years by intercalating a 13th month.

The earliest Roman calendar and the calendars used in the Greek states in classical times, i.e., after about 750 B.C., seem to have been kept in reasonably constant phase with the seasons by ad hoc intercalations of an extra month before the beginning of spring and were presumably seldom in error by more than 1 to 2 months and never by as much as a year.

The most consistently maintained ancient calendar appears to have been the Egyptian civil calendar, which had three 4-month seasons, making up 12 months of 30 days each and 5 days added at the end of the year to make a 365-day year, beginning with the season of Nile inundations.

Babylonian dates from a surviving list of kings and their reigns, copied by Ptolemy of Alexandria, with some fixes given by the recorded dates of eclipses, can be regarded as certain back to 650 B.C., and perhaps even 750 B.C. A list of the earlier Assyrian kings ruling in Mesopotamia, with cross-checks obtained from the correspondence they exchanged with Egyptian rulers, continues this chronology back to about 1350 B.C.

Ancient Chinese calendars, from the Shang dynasty (approximately 1766 to 1122 B.C.) were reckoned in cycles of 60 days, subdivided into 10-day periods, and combined into longer cycles of 12 solar years, which were fixed by observation of Jupiter (the length of a year on Jupiter is 11.86 Earth years).

Hindu chronology is counted from the death of Buddha, supposedly in 483 B.C.; but this date, from which ancient Indian documents are reckoned, is subject to an uncertainty of about 60 years, due to discrepancies in different sources, and may have been as early as 544 B.C.

Radiocarbon dating chronology of events 1,000 to 20,000 years ago are now quoted in years before 1952 B.P.

In conclusion, it may seem preposterous to devote a whole chapter to petty differences in dating in a treatise dealing with thousands and millions of years of Earth history. However, small differences in dating may involve hundreds and thousands of years when speaking of prehistoric times. Velikovsky[2] has corrected these differences and arrived at data reconciling the legends of the Bible, the Gilgamesh epos, and the Iliad with modern paleoclimatic records. Velikovsky's theories bridged a gap of 600 years and have aroused many controversies.

C. Age of the Earth

The earliest estimate not dependent on Biblical evidence was that of 100 million years suggested by Lord Kelvin in 1883,[3] on the basis of:

1. The rate at which the rotational speed of the Earth declined:
2. The time for which the energy output of the Sun could have been maintained
3. The time taken for the Earth to cool to its present temperature state from a molten condition

In 1900 Joly[4] produced supporting evidence for Kelvin's figure by considering the rate at which salt was being carried into the sea by rivers, and the time necessary for an initially "freshwater" ocean to acquire its present salt content. The discovery of radioactivity in 1896 and the subsequent realization that radioactivity is widespread in the rocks of the crust invalidated Kelvin's arguments and provided a "clock" by which rocks could be dated (*radioactive dating*). The oldest dated rocks (3900 million years old) are found in west Greenland; they consist of *anorthosites** which are included into an older series of rocks. These rocks, including the anorthosites, must be older than 100 million years or more. Thus it is almost certain that rocks exist with an age of greater than 4000 million years. The origin of the Earth, if it can be dated at all, must antedate the oldest rocks by some period at which we can only guess. Another method based on the relative abundance of the various isotopes of lead in galena (PbS) yields figures of 5000 to 5400 million years. Astronomical data suggest an age for the solar system of 5000 ± 1000 million years, which is in reasonable agreement with geochemical evidence. Rock samples from the moon have been dated at 4700 million years, also in good agreement with these estimates.

III. METHODS OF DATING

A. Radiometric Dating

Physical methods of dating, which depend on measurements of the residual radioactivity in matter in which this activity has been decaying since some definable time in the past, come first in precision of all methods available.

1. Radiocarbon Dating

This method claims accuracy for organic objects between 1,000 and 20,000 years old. Radiocarbon ages are conventionally quoted in years before present, meaning before A.D. 1952, about which time this method of dating first came into use.

The possibilities of using radiocarbon to date the past were first worked out by Libby,[5] based on the following reckoning: Atoms of the radioactive isotope of carbon ^{14}C are formed in the atmosphere of the Earth as a result of the continual bombardment by cosmic rays, mostly protons from elsewhere in the galaxy. The primary particles coming in vary widely in energy, but the peak frequency for particles from the galaxy is in the range of 10^8 to 10^{10} eV. Only those with energy exceeding 2×10^9 eV can pass through the whole atmosphere. Most are checked by collisions with atoms and molecules high in the atmosphere, splitting off free neutrons which are found most

---

*   Coarse-grained plutonic igneous rock.

abundantly in the stratrosphere. When these neutrons collide with atoms of nitrogen, the main constituent of the atmosphere, the dominant reaction produces an atom of radiocarbon: the atomic nucleus is changed by capturing the neutron and losing a proton (which appears as a free hydrogen atom). The carbon atoms produced in these ways sooner or later become oxidized ("burnt") and from then on form a minute part of the atmospheric carbon dioxide. Thus they become distributed throughout the atmosphere and oceans, entering into all the life processes in which CO_2 is involved, and are found throughout the biosphere and in the carbonates dissolved in the ocean, lakes, and in river waters deposited on their beds. The time required for the mixing through the different reservoirs of carbon represented by the atmosphere and the biosphere is evidently immaterial in relation to the average life of 8300 years before a ^{14}C atom disintegrates, for tests have repeatedly shown that the concentration of radiocarbon in living matter at any one time is closely the same all over the world.

A small proportion of all naturally occurring carbon consists of another isotope, ^{13}C, which is stable, i.e., not radioactive. The proportions of the different isotopes found in living wood today are $^{13}C/^{12}C \sim 1 \cdot 52 \times 10^{-2}$, $^{14}C/^{12}C \sim 1 \cdot 07 \times 10^2$. Any fractionation (i.e., differential transmission of the isotopes) occurring when the carbon passes from one reservoir to another or in the laboratory processes used in preparation of materials for dating measurements, must appear in a measurable anomaly of the ^{13}C concentration and the fractionation of the ^{14}C should be just double this. In practice, small corrections are found to be necessary and are applied on this basis.

Cosmic rays produce on the average two new radiocarbon atoms per 59 cm of the upper atmospheric surface of the Earth per second, and there is increasing evidence (from the radioactivity found in meteorites and in moon rock) that the flux of galactic cosmic rays concerned has remained constant over periods much longer than the life of radiocarbon. Hence, the ^{14}C proportion of atmospheric CO_2 is continually replenished, as is that in all vegetable matter that breathes it in and all animal life that feeds on this. And so, throughout the biosphere, the same proportion of ^{14}C in living matter is maintained.

The basis of radiocarbon dating is that from the moment an organism dies communication with the atmosphere ceases and the decay of the ^{14}C in it is uncompensated.

From measurements on ancient materials of known age, the best estimate of the half-life of radiocarbon has since 1962 been taken as 5730 ± 40 years.[6,7]* The half-life of tritium, being only 12.3 years, makes this product of the reactions with atmospheric nitrogen of no interest in connection with dating events in climatic history, though it is of interest in tracing atmospheric motions following the injections of artificially produced ^{14}C and ^3H resulting from nuclear bomb tests in recent years.

The experimental difficulties in obtaining meaningful and sufficiently sensitive measurements of ^{14}C for dating are formidable[5a,8] because of the abundance of other radiation, some of it high-energy particles, that has to be excluded from the counter. The radiation from the disintegrating ^{14}C atoms in the carbon dioxide gas carefully prepared from the sample of material to be dated consists of "soft", low-energy particles. Modern vegetation samples can be expected to give about 70 disintegrations per minute (dpm), samples of matter that lived 5730 years ago should therefore give 35 dpm, matter that is 11,500 years old about 17.5 dpm, and so on. The measurements are made on the carbon dioxide in the counter, itself built of radiation-clean materials,

---

* 5730 ± 40 years has since 1962 been accepted as the best estimate of the half-life of radiocarbon; but many dates have been explicitly calculated in terms of the figure earlier adopted, namely 5568 years, and about 3% should be added to radiocarbon ages using the earlier figure to make them comparable. Further revisions may yet be proposed, however, since one carefully conducted more recent measurement has suggested a half-life of 5833 ± 127 years.[6]

inside a protective hut built of lead bricks and lined with iron plates designed to cut out the radioactivity in the walls and floor of the laboratory. Yet even this hut is penetrated by the highly energetic cosmic rays which are capable of penetrating many meters thickness of rock, and the background count due to these amounts to 800/min. They are prevented from interfering with the measurement by a ring of counters surrounding the central counter, containing the carbon dioxide from the sample, which they switch off every time they register a cosmic ray; it remains switched off only for a fraction of a millisecond and is operative for about 99% of the time, and the count from the sample can be adjusted accordingly. A fairly constant rate of 13 counts per minute from background radiation of unknown origin gets through all these precautions. Counting is carried on for, say, 1000 min (16 hr, 40 min), sometimes longer when materials of great age are being studied.

Laboratories issue instructions to guide those gathering material to be dated,[8] particularly as regards the avoidance of contamination by carbonaceous materials of quite different age before, during, or after collection (clean polyethylene bags securely sealed are recommended). The quantity required is 5 g of actual carbon which is regarded as the minimum (corresponding to 40 g of charcoal, 75 g of wood, 100 to 150 g of peat, 300 g of shell, or 500 to 1000 g of bone).

Since the average life of a ^{14}C atom is 8300 years, a 1% error in the count means an error of 83 years. Because the decay is exponential, counts attributable to radiocarbon are detectable as an addition to the background with materials up to 50,000 to 70,000 years old at most, but the effects of contamination of the sample by older or younger matter introduce changes of count that give increasingly serious errors with advancing age: for a 0.1% contamination the error is about 300 years at 30,000 years true age and 3,000 years at 50,000 years true age. Material of infinite age with this contamination could appear to be only 57,000 years old.

The investigator collecting material for ^{14}C dating should develop a strategy for avoiding sites and samples liable to give false ages due to contamination *in situ* by older or younger carbon. Usually the best ancient materials for giving consistent and representative dates are charcoal, wood, and peat, the latter especially from raised bogs.

Various tests have provided satisfactory (indeed possibly surprising) evidence that carbon atoms, including the radioactive ones, do not migrate from year ring to year ring within a tree; though it is important to eliminate any mobile fluid (sap) from the sample to be dated, this is not in practice a source of significant error.

It is not sufficient to measure the deviation of the radioactivity in the sample of material to be dated from the level prevailing now or in 1950, because Man's activities in recent times have changed the concentration of ^{14}C in the atmosphere (and are still changing it) in addition to the continuing natural fluctuations on various time scales just discussed. The two chief causes of these man-made changes are

1. The burning of increasing rates of ancient carbon, in which all ^{14}C activity has long been dead, in coal, oil, gas, and all such "fossil fuels", has been increasing the amount of carbon dioxide in the atmosphere and diluting its radioactivity with "dead carbon". This industrial revolution effect, first pointed out by Suess in 1953, and since known as the "Suess effect"[9,9a] should have reduced the atmospheric ^{14}C by about 10% from the end of the 19th century to that time, were not some of this man-made CO_2 taken up by the biosphere and in the ocean. The observed dilution of atmospheric ^{14}C attributable to this cause seems to have been between 2 and 6% by 1940, i.e., perhaps half the man-made output remained in the atmosphere.

2. Nuclear bomb explosions produce so much radiocarbon that from 1954 onwards

the industrial dilution was substantially offset and in 1963 the amount of ^{14}C in the atmosphere, according to measurements made between 27° and 70°N, rose to twice (203%) the natural level.[10] By 1965, with a cessation of most weapon testing, the excess had fallen to 70 to 80% above normal. The 1963 ring in living trees is therefore clearly and easily identified by its ^{14}C activity, and any migration of ^{14}C atoms from it into the older wood should be easily traced.

Radiometric methods of dating rocks and other materials of high age are of great importance to geology in providing a time scale to the great spans of the past of the Earth which were hitherto only identified by name. Tree rings and varves may, of course, provide "floating chronologies" which could give some indication of the year-by-year weather sequence, at the place where they were found, over some period beyond the range of precise dating.

2. Uranium Series Dating

The various naturally occurring isotopes of uranium (^{238}U, ^{235}U, etc.) decay ultimately to lead, by passing through characteristic successions of atomic disintegrations forming elements of intermediate atomic structure and widely differing half-lives. In the disintegrations helium gas (^4_2He) is given off or trapped within the surrounding rock. Some of the elements produced have half-lives which make the quantity present of interest as a means of dating the substance in which they are found.

In theory the absolute amounts of any of these elements present, or the ratios of the quantities of different ones, may be used as an indicator of the time elapsed since the beginning of the process that produced them. Thus, assuming that the accumulation of lead and helium, the two stable products, began at the moment of crystallization of a rock from volcanic magma, measurements of the amounts of either of these elements present will give an indication of the age of the rock. But in the case of helium measurements this must be regarded as a minimum age, since some of the helium produced may well have escaped. Ages obtained by measurements of the lead may be too old, falsified by preexisting lead present in the molten volcanic material in some cases, e.g., if there was an influx of older rock material remelted in the volcanic episode that preceded final crystallization. Thus, the ratios of the quantities found of the various unstable radioactive elements produced in the uranium and thorium successions should give a more reliable indication of the length of time during which the production has been going on.

3. Protactinium-Ionium Dating

Of most interest for our studies is the *protactinium-ionium method*[11,12] which has been applied to dating the material in ocean sediments up to about 200,000 years old. This measures the ratio of the protactinium(^{231}Pa) to the thorium isotope (^{230}Th) which is known as ionium. At the short-range end of its applicability, where it is subject to a wider margin of uncertainty than the radiocarbon method, this method has given results in reasonable agreement with radiocarbon.

4. Potassium-Argon Dating

A very small proportion of the radioctive isotope of potassium (^{40}K) exists alongside the common, stable isotope (^{39}K) in all potassium-bearing rocks and minerals. ^{40}K disintegrates, producing argon (^{40}A) and some calcium (^{40}Ca). The half-life of ^{40}K is 1300 million years. This very long time required for half the ^{40}K to disintegrate makes the method applicable to dating some of the oldest rocks on the Earth, the decay-time again being measured from the time when the rocks crystallized. The method depends on measuring the quantity of argon trapped in the rock, and thence the K/A ratio, and is therefore subject to error — liable to give a minimum age rather than the true age — if some of the argon has escaped since. It is thought to be subject to an error margin of about 100,000 years in the case of the youngest rocks to which it has been applied, at ages of about a million years.[13]

5. Rubidium-Strontium Dating

This is another method applicable to rocks which do not contain the heavy metals of the uranium group. Rubidium is a metal in the same column of the periodic table as sodium and potassium. It occurs widely, though only in very small quantities, in two isotopes ^{87}Rb and ^{85}Rb. About 27% of it consists of the isotope ^{87}Rb, which decays very slowly to ^{87}Sr (strontium) by emitting a beta-particle. The half-life of ^{87}Rb is 53,000 million years. This still longer half-life than that of ^{40}K, as well as the very small quantities available to work with, mean that the method is only suitable for dating the oldest rocks and to estimate the age of the solar system itself.[14]

6. Oxygen Isotope Dating

In deciphering ancient climates, the most useful means are those which yield direct measurements of past temperatures, such as stable isotope methods. Thus far, oxygen isotopes have proved most useful and measurements of deuterium abundance are beginning to have application to paleoclimatology. The significance of carbon isotopes in these kinds of studies is obscured by factors which have little relationship to climates or which are not yet completely understood. There are no other methods for estimating paleotemperatures directly.

Limitations on oxygen isotopes include the assumptions which must be made about seawater composition and the absence of ice caps, as well as the restrictions due to alteration of the original rock material. In effect, the paleotemperature record from oxygen isotopes is meaningless for the interval since the middle Miocene and the record from carbonate rocks is of doubtful validity for preCretaceous times. Analyses of oxygen isotopes in cherts* eventually may extend the range of applicability of the technique back to the beginnings of the history of the Earth but presently available determinations are not abundant. Future measurements are not likely to be closely spaced in time because of the relative scarcity of chert in the geologic column.[40]

B. Amino-Acid Dating

Organic chemistry may provide a means of dating shells and other organic detritus in the stratigraphy of a sediment back to about the mid-Tertiary, perhaps 30 million years ago, through details of the amino acids present, as first hinted by Abelson.[15] These polymers, like proteins, are among the basic structures required by living organisms. Some amino acids appear stable and are present in similar amounts in recent as in Tertiary shells of the same species, whereas others are relatively unstable. Since the decay of the different amino acids depends only on temperature and time, and some have been studied in the laboratory at high temperatures (up to 225°C) where decay proceeds rapidly, it should be possible to deduce the decay rates applicable at the temperatures between 0 and 30°C prevailing on the Earth and apply the results to geochronology over millions of years. Very small quantities of the order of 100 mg suffice for these tests. Experience of the method using different substances and reactions is needed to indicate error margins.[16]

Bada et al.[17] have shown how the slow racemization of amino acids in human and animal bones can be used, subject to assumption of the overall average temperature experienced by such material at the site concerned, to date remains back to at least

---

*   Compact rocks consisting essentially of cryptocrystalline quartz.

120,000 years B.P. Alternatively, in cases where the dating can be established by radiocarbon — i.e., within about the last 40,000 years — the average temperature of the material over the time since deposition can be deduced wthin a few tenths of a degree from the observed progress of racemization. These possibilities arise because the change in the substance depends only on temperature and time. Dating beyond the limits of the radiocarbon method is possible because of the longer half-lives of amino acid reactions: the case of the racemization of aspartic acid and the production of D-alloisoleucine from L-isoleucine at 20°C. Radiocarbon age determinations on younger materials at the same site can be used to calibrate the rate of racemization. Using the aspartic acid reaction, dates of bones of cave dwelling men and animals at the coast of Cape Province, South Africa, associated with a life that required high sea level, appeared accurate within 10,000 years at 120,000 B.P., and 5,000 years around 65,000 B.P.

C. Thermoluminescent Dating

Some minerals, e.g., quartz, have the property of acquiring thermoluminescence by exposure to radiation, whether this be of natural or artificial origin. This property is due to the trapping of electrons in the crystal lattice. When the mineral is heated, these electrons escape and are seen as the emission of a glow. Heating above 500°C eliminates all this previously acquired accumulation of energy represented by the trapped electrons. Thus, measurements of the thermoluminescence of any substance when heated are a measure of the radiation dose it has accumulated since the last time it was heated above 500°C. The radiation dose accumulated depends on the nature of the mineral and the amount of radiation to which it has been exposed in the time elapsed. The rate at which the particular substance acquires thermoluminescence can be tested in the laboratory by exposure to radiation from an artificial radioisotope. This provides a method of dating the time at which volcanic rocks cooled or at which a given article of pottery was fired.[18]

D. Tree Ring Dating (Dendrochronology)

Yearly growth rings are prominent in the stems of trees growing in the seasonal climates of middle and higher latitudes. The newest cells, and therefore the youngest ring, are those next to the bark. Each tree ring is formed by the succession from larger cells and softer wood, produced in spring, to smaller cells and harder, denser wood, produced in summer and fall (Figure 1). Microscopic examination of sections from sufficiently sensitive types of tree, particularly if supplemented by X-ray photography for subsequent measurements of differences of optical density to be calibrated against physical density of the wood, can even reveal some detail of the succession of periods more or less favorable for growth within individual seasons.[19,20]

A climatic event, such as an extremely harsh season or a group of extreme seasons can be identified with a year or years of anomalous tree rings and thus be dated in favorable circumstances to the exact year. More usually, there is some margin of uncertainty due to occasional years in the tree's life which produced either a very thin ring or, alternatively, a double ring. When different trees in the same locality are examined, the thinnest rings produced by the worst years for growth may appear to be missing altogether in the sections from some trees. In areas where standard tree ring chronologies present themselves, the unresolvable uncertainty of dating of any feature rarely amounts to more than a year or two, even when the chronology extends back over several thousand years.

The longest tree ring chronologies in the world have been established for the very long living trees, the Giant Redwoods *(Sequoia)* and the Bristlecone pine *(Pinus aristata)* in the western U.S. In the case of the Bristlecone pine in the White Mountains

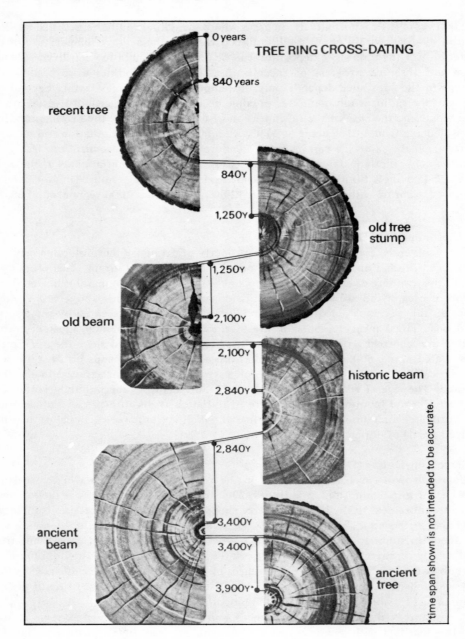

FIGURE 1.   Tree ring cross-dating. The diagram shows how the pattern of one set of tree rings can be added on to the rings of an older tree. The date of cutting of the trunks can be established with considerable accuracy and allows dating of additions or alterations to a building. (From Waechter, J., *Prehistoric Man*, Octopus Books, London, 1977. With permission. Illustrations by Ralph Stobart © Octopus Books Limited, originally published in *Prehistoric Man* by Octopus Books Limited.)

of east central California, the chronology of year rings extends back over 7000 years to before 5000 B.C. And in these best tree ring series repetition tests showed no error back at least as far as 3535 B.C.[21]

Lichenometry which uses dating by growth of lichens is a special method applied to dating the retreat of glaciers and disappearance of perennial snowbeds. Lichen spores reach the newly exposed rock faces, and some species (e.g., *Rhizocarpon geographi-*

cum) are thought to have nearly constant growth rates after an initial flush. Growth is slow and continues over a very long time. An estimate of the time elapsed since the rock became accessible to colonization by lichens can be obtained by measuring the size of the largest lichen present. However, errors may arise from growth rate changes due to subsequent climate variations.

E. Pollen Zone Dating
1. History

It has long been recognized that a sequence of past changes in the vegetation — in its composition at any given place and in the geographical limits of genera and species — is registered in the pollen found in layered sediments of lake beds and at different depths in peat and undisturbed soil and subsoil. This is the case in all parts of the world, including the Arctic tundra, the lowlands of the temperate zone, and the mountains in every continent. Moreover, a number of changes seem to have taken place suddenly, so that in any column of sediment or soil that may be examined distinct zones of different pollen assemblages — associated in some places with characteristic larger remains (e.g., tree stumps) and changes in the condition of the peat (e.g., dark, dense and full of humus where it was at some time exposed to a dry climate, much lighter and less humified where growth continued vigorously in a moist regime) — are more or less sharply marked off from each other. In Norway, the botanist Blytt[22] in 1876 took the prehistoric changes of vegetation which he recognized in the stratigraphy of peat bogs and lakes to identify a sequence of climatic changes that evidently divided up the history of postglacial times. At first these changes were dated just by their position in the stratigraphy. The dating was later progressively refined as new methods became available. After the first steps in pollen analysis around the turn of the century, Sernander[23] extended Blytt's scheme to Sweden, where cross reference to De Geer's [24,24a] varve chronology then became possible. An obviously related sequence of vegetation, and presumably climate, changes were soon seen to apply in other parts of Europe, and the Blytt-Sernander scheme began to be given much wider currency. Systematic pollen analysis and radiocarbon dating have since given much fuller knowledge of the distribution of the climate and vegetation changes and their timing (Figure 2).

Another feature of peat bog stratigraphy which has been used as a marker for dating are the recurrence surfaces.[7] These register renewed growth of the bog in a moist climate after a dry phase, by a change in color from dark to light.

The earliest evidence of life on Earth is the rock formation of stromatolites which were formed 3 billion years ago by microorganisms. Proof of such early microfossils has been found in western Australia by Schupf[25] of UCLA at the bottom of a shallow sea. Five fossil varieties of chains of cells, looking like tiny strings of beads, have been identified so far.

2. Advantages

Pollen analysis has proved in the past 50 years to be an excellent tool for reconstructing former vegetations. The method relies on the fact that the outer walls of pollen grains consist of a very resistant organic material. The pollen grain has a few properties that make it very useful for dating. It can, for instance, be preserved for many millions of years in sediments if protected against oxidation. Also, pollen grains are produced in enormous quantities by many species of plants. The pollen rain produced during the flowering of the plants will be distributed over a relatively large area and is to some extent related to the composition of the vegetation of that area. The outer wall of the pollen grain shows intricate specific patterns of ornamentation. Therefore, it is possible to identify the plant from which the preserved remains of a pollen grain originated, sometimes even at the specific or lower level. Another advantage is the small-

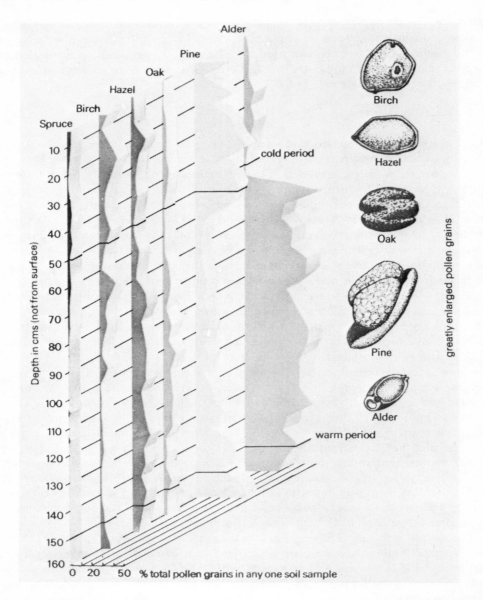

FIGURE 2.   Palynogram and pollen rain. In a cold period conifers (e.g., pine) thrive and deciduous trees (e.g., alders) decrease, resulting in a higher percentage of pollen from the former type. The reverse happens in a warm climate. As pollen grains are virtually indestructible, they indicate whether warm or cool conditions prevailed at any given time. (The Palynogram for Israel has been described by Horowitz, A., *Quadmoniot*, 13, 80, 1980). (From Waechter, J., *Prehistoric Man*, Octopus Books, London, 1977. With permission. Illustrations by Ralph Stobart © Octopus Books Limited, originally published in *Prehistoric Man* by Octopus Books Limited.)

ness of the pollen grain (10 to 150 μm), which makes it possible to extract even from small samples a sufficient number of individual grains to permit quantitative results. This enables narrow sampling intervals, sometimes 1 cm or less in suitable sediments, and makes it possible to distinguish changes in the pollen rain, and consequently changes in vegetation, within time intervals as short as 1 year.

3. Palynogram

Pollen analysis can contribute to the solution of various problems in different areas of the historical sciences. Pollen-paleological (palynological) problems might involve the history of vegetation, the climatic history of a region, the correlation of sedimentary sequences, the subdivision of a plant, and the effect of Man on nature. The result of a palynological survey can give us some insight into principles and concepts pertaining to organization at community level and also into the development and evolution of former ecosystems. The results of pollen analyses are presented in pollen diagrams, a graphic presentation of the percentages of individual pollen types. In order to summarize the results in a way that better illustrates the observed changes in percentages, general diagrams are constructed. In these diagrams, groups of pollen types, belonging to plants of a certain group, are taken together and so it is possible to observe the changes from a tundra to the northern coniferous world, or from a temperate deciduous forest to temperate grassland (Figure 1).

4. Pollen Rain

The basic concept in palynological work is the "pollen rain", i.e., the mixture of a pollen that comes into circulation and only that part that is fossilized in the sedimentary sequence. The composition of the "pollen rain" is dependent on several factors: first, the frequency of the species in the region; second, on the absolute pollen production of the species. This production itself can vary specifically and individually according to the conditions under which the specimen grows — in the middle or at the boundary of its geographical area, in an open position, or in a closed stand.

The frequency of flowering years is also important. There are forest trees which flower only at intervals longer than 1 year because sometimes a period of rest is needed before a species can flower profusely. The nearer one comes to the limit of the area of distribution, the longer the periods of rest between flowering. The dispersal mechanism of the pollen is also important; dry pollen is better dispersed than pollen that sticks together. Moreover, pollen production is not the same for every plant, so the "pollen rain" must be corrected for the differences in pollen production. This problem is solved by comparison with the "pollen rain" of recent vegetation. As long as the forest is composed of trees that produce comparable amounts of pollen, the actual composition of the forest is comparatively well reproduced. This is the case in northern Finland. The situation is more complicated if the main forest trees are seriously underrepresented as is the case in the tropical rain forest.

5. Evaluation

With regard to climatic change, changes in "pollen rain" can be translated into changes in climate. The remains of different species are found in samples analyzed for pollen grains: these include the pollen of higher plants, spores of mosses and lycopodes, and vegetative and generative remains of fungi and algae. In general it can be stated that the last categories give us a different type of information from the first ones. From fluctuations in the first categories of remains, information is implied concerning the relation between regional vegetation and climate. The last categories give us more information on local topographic circumstances. So we must be very critical in evaluating the information given by the various components of our record.

The problem is to find a method that extracts climatic evidence from the ecological data which indicate the components of the vegetation present according to the palynological evidence. There are two general ways of approach: (1) "autecological" or (2) "synecological".

In the "autecological" approach the climate is reconstructed by a detailed knowledge of the ecology of certain species. For instance, it is well known that water plants have several features which make them especially valuable for ascertaining late glacial temperatures, because they have a higher dispersal rate, partly because they are transported by migrating water birds over long distances, and also because they do not require any specific soil conditions before they can thrive. So a good knowledge of the ecology of aquatics will give us valuable information on temperatures. The ecology of land plants is, however, also useful.

The second "synecological" approach to the transfer of floristic data into climatological parameters has been chiefly developed in America. This method aims at the formulation of mathematical transfer functions; by means of these one can translate records of fossil pollen directly into quantitative estimates of climatic parameters. In this method a data base must first be constructed. In this data base all possible types of recent regional vegetations are incorporated with their "pollen rain", together with their climatic data, such as temperature, rainfall, presence of air masses, and so on. The assumption is made that a fossil "pollen rain" is comparable to recent vegetation, as present in our data bank. Then by comparing the recent vegetation with its climate, an impression of the fossil climate is obtained. This approach requires several assumptions:

1.   The observed floristic composition of types, as assessed by the samples of recent pollen in a certain area, is in balance with the moderate climate.
2.   The climate is the ultimate cause of changes of pollen records in the time.
3.   The climatic response exhibited by each pollen type has remained constant during the time interval under discussion.
4.   The relation between each climatic variable and a set of pollen types is linear.

Assumption 1 is fulfilled during the interglacials, but may not be fulfilled during the interglacial-glacial transitions and during the interstadials because the transition time or the duration of an interstadial was too short.

A complication arises when the upheaval of a mountain chain has created new areas for occupation. During such a process of local climatic changes and plant immigration new species are created and preexisting plants may respond to climatic variation in other ranges of their ecological amplitudes, so that new response functions become initiated and different transfer functions must be used. Assumption 2 is very often valid and is only not fulfilled during a transgression caused by isostatic readjustment and in newly originated habitats with a fast rate of evolution. Assumption 3 is correct in the middle range of the ecological amplitude of a plant but when conditions are marginal, unusual responses may occur; an example is the profuse flowering of plants when ecological conditions are extremely critical.

6. Reconstructing Ecosystems

One way of reconstructing former climates and therefore of recognizing climatic change is the reconstruction of former biota (flora and fauna). This can be done by analyzing those parts of these biota that fossilize in suitable environments. From the fact that it is not possible to cultivate the same crops everywhere we can conclude at a glance that a specific life form (plant or animal) makes certain demands on its environment. This reasoning may be reversed — the presence of a life form for a certain time span means that the requirements of that plant or animal as to its environment were fulfilled. By this simple argument it is possible to demonstrate that a plant or animal species can be seen as a mapping from the abiotic world into the biotic world. By analyzing the function or functions which define this mapping it is possible to reconstruct, from the presence of a certain combination of plants and animals, the abiotic world that accompanies this particular ecosystem.

An ecosystem is an ecological entity composed of plants, animals, and microorganisms and accompanying factors, such as temperature, light, precipitation, evaporation, nutrients, and so on. Thus, ecosystem can be defined as a subset of the biosphere in which the elements are in constant relation to each other. But the ecosystem is not a closed system; solar radiation and precipitation are external inputs, and an exchange of carbon dioxide, oxygen, and water takes place between parts of the ecosystem and the atmosphere by convection and wind. At the same time, in humid areas, a part of the precipitation is delivered to the ground water and another part leaves the system as runoff. Part of the incoming radiation is reflected and minerals are added by sedimentation of dust and so on, while exchange of organic material takes place through the movements of animals.

The effect of regional differences in climate on an ecosystem is best seen in the middle European forest system. In this system, beech, oak, and hornbeam are in competition with each other. As the climate becomes warmer and drier, or more humid, the beech is replaced by different oak species. The oak, at increasing continentality, is in turn replaced by the hornbeam. In this system the growth conditions of plants belonging to the under story are dependent on the composition of the upper story. This is not always the case. In a peat bog, for instance, a soft moss *(Sphagnum)* cover is the guiding element. Besides *Sphagnum,* only those species that, even under poor nutrient conditions, produce so much matter that they are not overgrown by the *Sphagnum* will survive. So in a wet oceanic climate the growth of *Sphagnum* is so intensive that pine cannot maintain itself, while in more continental parts the growth of *Sphagnum* is inhibited and the peat bog becomes more and more overgrown by pine. This example gives us an idea of how different climatic conditions and precipitation lead to a different vegetation type which will also show up in a different way in the fossil record.

A close correlation exists between plant cover and climate; if we can reconstruct former vegetations, we have a concealed indication of former climates, as long as we can use plant species that are not extinct. We then make the assumption that the climatic requirements of a plant species do not change with time.

7. Summary
From the evidence brought forward in this chapter it must be concluded that, in general, the great fluctuations in climate were contemporaneous in the southern and northern hemispheres. Secondly, the advances of the glaciers, coinciding with the extreme phases of the glacials, were contemporaneous with the generally drier climate of the Earth. The only place where an indication of a wetter climate than now was encountered is the area of the southwest of the U.S. and possibly an area south of the Atlas mountains. Within glacial periods, ameliorations of climate occurred, called interstadials. During these phases a rise in temperature, accompanied by increasing precipitation, took place. The wetter phases, formerly called pluvials, do not coincide with the extreme phases of the glacials, but occur, in general, at the transitions from glacial to interglacial conditions. Within the period from roughly 50,000 to 22,000 B.P., a general rise in temperature took place, accompanied by an increase in precipitation.[38]

F. Lake Varves Dating
Mud and clay sediments formed at the bottom of lakes (found also where former lakes have dried up and been buried by a later soil and vegetation), and similarly formed on some parts of the sea bed near land, commonly show yearly layers called "varves". These are produced by regularly alternating sequences of color, grain size, and mineral and organic composition, due to seasonal differences in the relative amounts (and kinds) of organic material and wind- and waterborne mineral particles deposited. These differences are particularly pronounced where the streams feeding a

lake are frozen for part of the year and the seasonal thaw is followed by a period of rapid or torrential flow. Wherever a thicker or thinner varve is attributable to some climatic event, this can be dated in the chronology represented by a yearly varve series, much as with tree rings.* As a method of dating, the study of varve series was pioneered by De Geer[24,24a] and used as a first means of dating the retreat of the edge of the great Scandinavian ice sheet across northern Europe at the end of the last glaciation.

G. Dating of Volcanic Layers (Tephrochronology)

Within the fan-shaped area of deep deposit of debris (tephra) from a single great volcanic eruption, or the wider region covered deeply by layers of ash in a period of abnormal activity of the nearer volcanoes, the age of which has been established by radiometric dating or other means, this layer provides a fixed point in the dating of the stratigraphy wherever it can be identified. In some cases, a whole sequence of tephra layers of known date provides a fairly recognizable chronology, though of no more than regional significance.

Probably the most extensive scale on which this method has been used is seen in the series of postglacial tephra layers found by Auer[27] first in Tierra del Fuego and later over wide areas of Patagonia and used by him to date the stratigraphy of many peat bogs. The chronology and the series of waves of heightened activity of the south Andean volcanoes which it represents was specified by Auer as follows:

Layer Number I — A stratum of pumice particles, apparently 17,000 to 16,000 years old.

Layer Number II — Another stratum of pumice particles thought to be 14,000 to 13,500 years old.

Layer Number III — An ash layer, apparently about 12,000 years old.

Layer Number IV — Another ash layer, thought to be 11,000 years old.

Layer Number V — A white rice-like tephra layer, commonly 1- to 3-cm thick, ^{14}C dated as about 9000 years old.

Layer Number VI — A greenish, or yellow-grey, brown tephra layer, 5- to 30-cm thick, the coarsest grained of all the postglacial tephra layers in Patagonia, chemically similar to earlier layers in Late Glacial times. This layer was ^{14}C dated as 5500 to 5000 years old.

Layer Number VII — A white tephra layer, commonly 3- to 10-cm thick, ^{14}C dated as 2500 to 2200 years old.

Layer Number VIII — Light gray ash forming a stratum at or near the top of the pollen stratigraphy profiles of the Patagonian peat bogs. From the range of depths concerned, in relation to the depths of the ^{14}C dated layers I to III lower in the bogs, this ash layer must represent volcanic activity in the last 400 to 500 years. Other countries for which such chronologies of volcanic deposit layers can be established include Iceland, Japan, and New Zealand.

Similarly, the great eruption of Santorin (Thera) in the Aegean about 1500 and 1450 B.C. produced a thick ash layer over the regions around the eastern Mediterranean.

Dating of volcanic layers themselves has usually been determined from ^{14}C dating tests performed on organic carbon — humus from peat or the remains of marine organisms on the sea bottom — immediately above and below the volcanic material. In the case of ash buried in an ice sheet, the age may be determined either by the stratigraphy of annual ice layers or by the flow method described above. In some parts of

---

* Not all varve series represent an annual cycle of deposition. Some in small lakes are produced by much shorter time scale variations in the material deposited. But the conditions producing varve formation have been studied, and at least in "large" lakes it seems that the yearly rhythm predominates.[26]

the world extensive deposits of thick strata of volcanic material, bearing witness to long periods in the past of much greater volcanic activity than now, can be roughly dated by their relationship to the magnetic reversals and their place in the stratigraphy.

H. Ice Sheet Dating
1. Stratigraphy
Recognition of yearly layers in an ice sheet is made possible by the similar sequences of grain size, hardness, and density, normally culminating in an autumn layer which is hard and of high density, overlying the coarser grained summer layer which may or may not show evidence of melt processes. The autumn layer is sometimes a single wind slab or it may be a succession of closely spaced wind crusts. It is formed as a result of the strong vertical gradient of temperature within the snow when the surface cools, causing upward transfer of vapor in the air trapped within the snow, leading to recrystallization at the surface; the upward transfer is hastened by the turbulence produced by winds blowing over the snow surface, which remains colder than the summer layer beneath.[28] The amount of the annual accumulation is established by weighing. The strata remain recognizable, despite compression by the ice above, for many decades.

2. Flow Method of Dating Ice Sheets
Striking successes seem to have been achieved in dating the ice below the levels where all usable traces of the year layers become obliterated by molecular diffusion in both the Greenland and Antarctic ice sheets by applying a simple concept of the flow of the ice.[29-30]

Verification of the method as probably giving some approximation to real ages, and the justification thereby of the daring assumptions on which the theoretical derivation of the ages rests, may be regarded as indicated by the results. These have entailed a further implied discovery, which could not have been taken for granted in advance, namely that the sequence of changes of prevailing temperature given by oxygen isotope analyses of the ice substance points to about the same dates in Antarctica and Greenland for the onset (around 70,000 years ago), culmination (between 17,000 and 20,000 years ago), and ending (around 11,000 years ago) of the cold climates of the last Ice Age as are indicated by most evidence in Europe and elsewhere. If we may assume that the dates should really coincide and that they correspond (as they seem to do) to global changes of climate and atmospheric circulation taking place more or less abruptly over the whole Earth, then the error margin attaching to dates of the ice derived from the flow model cannot be more than 5% and may be very much less.

I. Ocean Sedimentation Dating
It is now understood that the ocean bed is a dynamic environment. Slumping of sediments on slopes and drift and scouring of bottom deposits by water currents occur in many places. Moreover, in such cases the land-derived clay particles and the carbonates deposited as the shells and bones of marine organisms are not only of different origin but are affected differently as regards solution and transport so long as they are exposed to the action of the water. Too little account still seems to be taken of the fact that runoff from the land, and the supply of riverborne minerals and organic matter, and also the rate of circulation of the ocean water itself, must have changed significantly as ice ages succeeded warm eras and vice versa. For these reasons, the best sites for sediment examination are probably on rises of the ocean floor in the deep regions well away from the continental slopes. In such places, according to Ericson and Wollin,[11] the rate of deposition of the sediment has remained reasonably constant through all the great climatic changes in the Pleistocene. Nevertheless, the rapid sedimentation of deposits on the continental shelves must offer a more detailed record. They continue to attract interest accordingly.

The composition of the flora and fauna in the upper oceanic waters largely depends on its temperature, although such influences as salinity and predators may cause local differences. As these organisms die and accumulate at the bottom of the ocean, the fossil content of the deep sea sediments will contain a filtered reflection of the surface composition because of partial or total solution of certain tests as they sink or lie on the bottom. The recognition of a particular species, such as *Globorotalia menardii,* within the sediments can therefore be a strong indication of warm water at the ocean surface at that time, while its absence from the microfossil content will suggest cold conditions. Possibly better quantitative temperature estimates can be obtained by studying the total fauna whereby small changes in the fossil community can be used to indicate relatively small temperature fluctuations which would not be so well reflected by examination for the relative concentration of only a few species.[39]

J. Geomagnetic Dating
1. Geomagnetic Rules
The magnetic field of the Earth undergoes variations of strength, which are usually assumed to be cyclic with a period of the order of 10^4 years between the minima; Bucha[31] gives some details for the last 10,000 to 20,000 years. At usually much longer intervals, the main dipole magnetic field of the Earth reverses its direction and is presumably very weak for some time during the reversal. A fossil record of the reversals is preserved in the remanent magnetization of the rocks, imprinted in them at the time of crystallization, as may be shown by heating rock or pottery and allowing it to cool in a magnetic field. The reversals are registered in sediments containing volcanic materials, e.g., ash or lava. There is no apparent regularity or periodicity in the timing of the reversals, of which very many are recorded in the history of the Earth and at least 27 in the last 4½ million years, the total duration of either polarity being identical, and the mean field strength in either case, too. "Epochs" of dominance of either "normal" (as now) or reversed polarity seem most frequently to last about a million years; the longest "epoch" recognized spanned 5×10^7 years, the shortest under 10^5 years. There are, however, also shorter episodes (known as "events") of opposite polarity, typically lasting a few tens of thousands of years within the longer "epochs".

2. Geomagnetic Changes
Reversals of the geomagnetic field have occurred on the average at least three times per million years for the last 70 million years. The change of polarity of the geomagnetic field seems to take some 2000 years, but is preceded and followed by periods of some 3000 years of reduced intensity of the field. Studies of such reversals in deep sea sediment cores indicate that the reversals seem to coincide with periods when index fossils tend to appear or become extinct and it was suggested that this may be due to changes in the protectiveness of the Van Allen radiation belts against incoming, harmful radiation during the period of a weak geomagnetic field. This mechanism is thought to be inadequate in its direct effect but may operate by causing climatic change.[39]

3. Archaeomagnetic Dating
The magnetic reversals may also be marked by extinctions of species of fauna, since the surface of the Earth is presumably subject to an increase in cosmic ray bombardment when the geomagnetic field is weak, and this should cause an increase in the incidence of mutations: evidence of such effects in the minute marine fauna represented in ocean bed sediments has been reported.[32,33] It has been suggested that times of weakness of the magnetic field of the Earth may be times of intense volcanic activity and that, through this, they may tend to induce ice ages. It may be possible to settle this, if dating techniques become precise enough to establish whether there is indeed a correlation between the reversals and ice ages.

The shorter time scale variations of the magnetic field of the Earth may also be useful in dating, known as archaeomagnetic dating because of its application to the times with which archaeology is concerned.

K. Archaeological Dating
1. Human Progress

Man-made things — known as human artifacts — ranging from highly artistic pottery and metal ornaments to the most primitive bone tools from Paleolithic times, which can be dated by radiocarbon tests or other independent means, have enabled archaeologists to build up a knowledge of the history of the discovery of new materials and methods of working, and of artistic styles, as well as of the places and times of origin of each. Wherever such objects are found buried in undisturbed soil or subsoil layers, identification of the culture to which the object or style belonged sets an approximate date, or more reliably, a "maximum" age for that point in the stratigraphy. This has been of use chiefly in increasing knowledge of the distribution of trade and communications in prehistoric times and of the time lags involved in the spread of fashion and new techniques, rather than in dating the marks of climatic history; however, it may supply useful corroboration of the latter in some cases. Moreover, even the trade routes used — for instance, when they crossed the Alps, the Sahara desert, or a great sea — may have interesting climatic implications.

The following method of dating for other important discoveries and advances in technique has been compiled by Childe[34] and Clark and Piggott:[35] flint, bone, tools, arrows, boats, pottery, copper (3500 B.C.), pure tin (1500 B.C.), bronze (3000 B.C.), and iron (1200 B.C.).

2. Domestication of Animals

Animal husbandry seems to have progressed gradually. The people of the earliest towns and villages near Jericho and in southern Turkey and Iran bred sheep and goats by around 7000 to 6000 B.C., but not yet cows. By 3000 B.C. the peasants all over the Near East, from Egypt to Mesopotamia, had yoked oxen to the plough. And by about 2500 B.C., if not earlier, asses and horses were harnessed to two-wheeled carts in Hittite, Asia. The first use of wheeled vehicles and horses was very early in the Neolithic epoch in Poland, i.e., also by 2500 B.C. or earlier. *Sledges* are known to have been used from quite early postglacial times, and by 2000 B.C. heavy sledges up to 4-m long, drawn by dogs, were in use from Finland to the Urals. Reindeer sledges had reached the Baltic by 1000 B.C. *Skis* go back to Stone Age times in Norway and are clearly shown in Karelian rock drawings between 3000 and 2000 B.C. *The beginnings of agriculture* go back to perhaps 9000 B.C. in those areas of the mountains in the Near East where the grasses from which modern cereals have been developed grow wild.

3. Fossil Man

Many discoveries have been made concerning Man's ancestry, especially in the last 10 years, and on the basis of these discoveries a detailed picture of the evolution of man has been compiled. Up to now only isolated discoveries had been made and no composite picture could be established (Figure 3).

The earliest hominids probably lived in Africa and were thought to derive from the bipedal savanna-dwelling anthropoid apes, like Ramapithecus. However, two main hominid groups were already present during the late Pliocene/early Pleistocene, the primitive Australopithecinae, and the more advanced forms representing the genus *Homo*. The Australopithecinae may well have been in direct competition with the early true man such as *Homo habilis,* but although present in S. Africa for a period of possibly 500,000 years they apparently gave rise to no descendants.

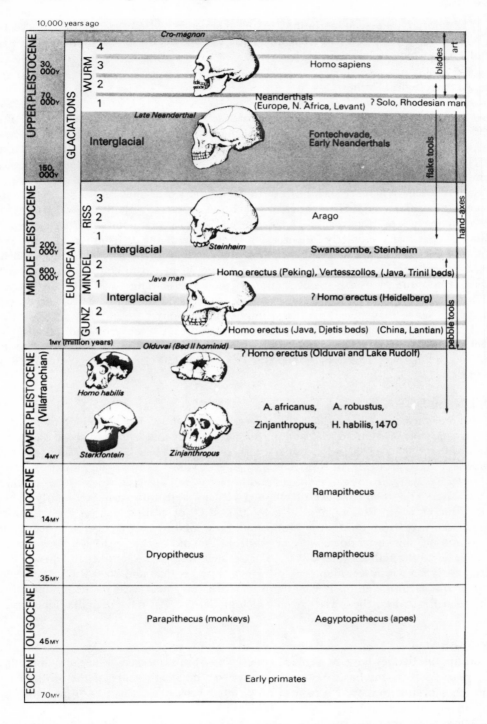

FIGURE 3.   Man's ancestry. Development of primates during the last 30 million years — the Cenozoic period. The first stage, the Eocene, saw the primates already established. By the second stage, the Oligocene apes and monkeys evolved among separate stems. In the Miocene, Man emerged with the Ramapithecus being the founder-member of the human line throughout the Pliocene. The Lower Pleistocene saw the advent of *Homo erectus.* The Middle Pleistocene is the main period of human development, followed in the Upper Pleistocene by Neanderthal Man and finally by modern Man *(Homo sapiens).* (From Waechter, J., *Prehistoric Man,,* Octopus Books, London, 1977. With permission. Illustrations by Ralph Stobart © Octopus Books Limited, originally published in *Prehistoric Man* by Octopus Books Limited.)

The genus *Homo* diverged rapidly, with *H. habilis* evolving into a subspecies of *H. erectus* as Java and Peking man together with African and European forms of Chellean man. It was the species *H. erectus* which finally gave rise to modern man (*H. sapiens sapiens*) via Swanscombe and Neanderthal man (*H. sapiens neanderthalensis*). There is considerable evidence that *H. sapiens sapiens* and *H. sapiens neanderthalensis* co-existed and that many of the physiognomic features exhibited by the latter are still to be found in some races of *H. sapiens sapiens*.

L. Astronomical Dating

The cyclic variation of the orbital elements of the Earth described by Milankovitch[36] must affect the amount of solar radiation available at different latitudes and seasons and to a much lesser extent the total for the year, as calculated by Vernekar.[37] There are good grounds for thinking that this must induce an alternation of warm and cold climatic periods on the Earth and that the timing of the last major cycle fits that of the last glaciation. If one accepts this as the cause, at least of the timing, of all the Quaternary glaciations, it provides a time scale which all fossil evidence of these ice ages should fit.

REFERENCES

1. **Huerzeler, J. and Schaefer, H.,** Der Mensch in Raum und Zeit, *Veroeff. Nat. His. Mus. Basel,* 1, 24, 1960.
2. **Velikovsky, I.,** *Ages in Chaos,* Vol. 1-5, Doubleday, Garden City, N.Y., 1952—1979.
3. **Kelvin, W. T.,** Age of the earth (1883), in *Climate: Past, Present & Future,* Lamb, H. H., Ed., Vol. 2, Methuen, New York, 1977, 835.
4. **Joly, M.,** Age of the earth 1900, in *Dictionary of Geology,* Whitten, D. G. A. and Brooks, J. R. V., Eds., Penguin Books, London, 1979, 17.
5. **Libby, W. F.,** Atmospheric helium-3 and radiocarbon from cosmic radiation, *Phys. Rev.,* 69, 671, 1946.
5a. **Libby, W. F.,** Radiocarbon dating, *Philos. Trans. R. Soc. London Ser. A.,* 269, 1, 1970.
6. **Berger, R.,** Ancient Egyptian radiocarbon chronology, *Philos. Trans. R. Soc. London Ser. A.,* 269, 23, 1970.
7. **Godwin, H.,** Radiocarbon dating, 5th Int. Congr., *Nature (London),* 195, 934, 1962.
8. **Duplessy, J.-C.,** Isotope studies, in *Climatic Change,* Gribbin, J., Ed., Cambridge University Press, London, 1978, 46.
9. **Suess, H. E.,** The three causes of the secular C_{14} fluctuations and their time constants, in *Radiocarbon Variations and Absolute Chronology,* Proc. 12th Nobel Symp. Uppsala, Olsson, I. V., Ed., Almqvist & Wiksell, Stockholm, 1970, 595.
9a. **Suess, H. E.,** Natural radiocarbon evidence bearing on climate changes, *CNRS,* 219, 311, 1974.
10. **Nydal, R.,** Variation in C^{14} concentration in the atmosphere during the last several years, *Tellus,* 18, 271, 1966.
11. **Ericson, D. B. and Wollin, G.,** *The Deep and the Past,* Jonathan Cape, London, 1966.
12. **Rona, E. and Emiliani, C.,** Absolute dating of Caribbean cores, *Science,* 163, 66, 1969.
13. **Fleck, R. J., Mercer, J. H., Nairn, A. E. M., and Peterson, D. N.,** Chronology of late Pliocene and early Pleistocene glacial and magnetic events in southern Argentina, *Earth Planet. Sci. Lett.,* 16, 15, 1972.
14. **Michael, H. N. and Ralph, E. K.,** *Dating Techniques for the Archeologist,* MASCA Handbook, MIT Press, Cambridge, 1971.
15. **Abelson, A.,** Paleobiochemistry, *Sci. Am.,* 195, 89, 1956.
16. **Hare, P. E.,** Geochemistry of proteins, peptides and amino acids, in *Organic Geochemistry,* Eglinton, G. and Murphy, M. T. J., Eds., Springer-Verlag, New York, 1969, 438.
17. **Bada, J. L., Protsch, R., and Schroeder, R. A.,** The racemization reaction of isoleucine used as a palaeotemperature indicator, *Nature (London),* 241, 394, 1973.
18. **Seeley, M. -A.,** Thermoluminescent dating in its application to archaeology, *Arch. Sci.,* 2, 17, 1975.

19. **Fletcher, J. M. and Hughes, J. R.,** Uses of X-rays for density determinations and dendro-chronology, *Bull. Fac. For.,* 7, 41, 1970.
20. **Le Roy Ladurie, E.,** *Times of Feast, Times of Famine,* George Allen & Unwin, London, 1972.
21. **La Marche, V. C. and Harlan, T. P.,** Accuracy of tree ring dating of bristlecone pine for calibration of the radiocarbon time scale, *J. Geophys. Res.,* 78, 8849, 1973.
22. **Blytt, A.,** *Essay on the Immigration of the Norwegian Flora,* Christiania, Oslo, 1876.
23. **Sernander, R.,** Die schwedischen Torfmoore als Zeugen postglazialer Klimaschwankungen, Geol. Foeren, Stockholm, 1910.
24. **De Geer, G.,** A geochronology of the last 12,000 years, Congr. Géol. Int., Stockholm, *C.R.,* 11, 241, 1912.
24a. **De Geer, G.,** Geochronologia suecica Principles, *Kgl. Vet. Akad. Handlingar — III,* 18, 6, 1940.
25. **Schupf, J. W.,** personal communication, 1980.
26. **Tauber, H.,** The Scandinavian varve chronology and C_{14} dating, in *Radiocarbon Variations and Absolute Chronology,* Proc. 12th Nobel Symp., Uppsala 1969, John Wiley & Sons, New York, 1970, 173.
27. **Auer, V.,** The Pleistocene of Fuego-Patagonia, Vol. 1-4, Suomal. Tiedeakat, Toim, Helsinki; *Geologica-Geographica,* I: 45, 226, 1956, II: 50, 239, 1958, III: 60, 247, 1959, IV: 80, 160, 1965.
28. **Benson, C.,** Polar regions snow cover, in *Physics of Ice and Snow,* Proc. Int. Conf. Low Temperature Sciences, Vol. 1, Part 2, Hokkaido University Institute on Low Temperature Sciences, Hokkaido, 1967, 1039.
29. **Dansgaard, W. and Johnsen, S. J.,** A flow model and a time scale for the ice core from Camp Century, Greenland, *J. Glaciol.,* 8, 215, 1969.
29a. **Dansgaard, W., Johnsen, S. J., Møller, J., and Langway, C. C.,** One thousand centuries of climate record from Camp Century on the Greenland ice sheet, *Science,* 166, 377, 1969.
30. **Epstein, S., Sharp, R. P., and Gow, A. J.,** Antarctic ice sheet: stable isotope analyses of Byrd Station cores and interhemispheric climate implications, *Science,* 168, 1570, 1970.
31. **Bucha, V.,** Evidence for changes in the Earth's magnetic field in density, *Philos. Trans. R. Soc. London Ser. A.,* 269, 47, 1970.
32. **Glass, B., Ericson, D. B., Heezen, B. C., Opdyke, N. D., and Glass, J. A.,** Geomagnetic reversals and Pleistocene chronology, *Nature (London),* 216, 437, 1967.
33. **Kennett, J. P. and Watkins, N. D.,** Geomagnetic polarity change, volcanic maxima and faunal extinction in the South Pacific, *Nature (London),* 227, 930, 1970.
34. **Childe, V. G.,** *What Happened in History,* 2nd ed., Penguin Books, London, 1954.
34a. **Childe, V. G.,** *The Prehistory of European Society,* Penguin Books, London, 1965.
35. **Clark, G. and Piggott, S.,** *Prehistoric Societies,* Penguin Books, London, 1965.
36. **Milankovitch, M.,** Mathematische Klimalehre, *Hbch. der Klimatologie,* Koeppen, W., Geiger, R., Eds., Borntraeger, Berlin, 1930, Vol. 1, Part 1.
37. **Vernekar, A. D.,** Research on the Theory of Climate Vol. 2, Long Period Global Variations of Incoming Solar Radiation, Travelers' Research Center, Contract No. E 22 — 137 — 67 N, Hartford, 1968.
38. **Wijmstra, T. A.,** Paleobotany and climatic changes, in *Climatic Change,* Gribbin, J., Ed., Cambridge University Press, London, 1978, 25.
39. **Tarling, D. H.,** The geological-geophysical framework of ice ages, in *Climatic Change,* Gribbin, J., Ed., Cambridge University Press, London, 1978, 3.
40. **Frakes, L. A.,** *Climates Throughout Geologic Time,* Elsevier, Amsterdam, 1979.

Chapter 4

THEORIES OF LONG-TERM CLIMATE CHANGES

I. CAUSES OF CLIMATIC FLUCTUATIONS

A. Basic Facts

Through long stretches of geological time, in periods before the Quaternary, the Earth seems to have had only warm climates with no great ice sheets; it is now thought, however, that there may have been many periods when bigger landmasses near the pole bore a cover of "permanent" ice. Indeed, this may have been true in most eras.

During the warm ice-free eras there would still be a general thermal gradient from the equator to the polar regions, though the overall temperature range at the surface on a yearly average was probably only one half to three fifths of the present one, and in midwinter a little under one half. In the upper air, owing to the presumably much weaker surface temperature inversions over the polar regions, the ratios near the 10-km level should have been as much as 80 to 85% of the present equator-pole range, on a yearly average, and 65 to 70% in midwinter. If at any epoch the solar radiation reaching the surface of the Earth and lower atmosphere were stronger than now, that would increase the equator-pole temperature difference, producing somewhat higher percentages of the present range than those suggested here; but the intensity of the present range depends greatly on the high albedo of the persistent snow and ice in and near the polar caps.

B. Changing Conditions

Fluctuations of climate are most conveniently considered in different classes according to the time scales involved which correspond to the action of different processes and different underlying causes. We may define the following classes:

1. Changes in the major geography, i.e., the crustal drift of the Earth (continental drift and pole wandering) and development, changes in the positions and heights of mountain ranges and the rate of rotation periods in the Earth in the range 10^9 to 10^7 years.
2. Changes due to the development of the sun (luminosity waxing) and fluctuations of solar output induced by rotations of the galaxy: time scales 10^9 to 10^8 years.
3. Changes of composition and radiative characteristics of the developing atmosphere of the Earth over the same time ranges.
4. Climatic fluctuations between glacial and interglacial regimes on time scales of the same order as the astronomical variations shown in the orbital arrangements of the Earth, i.e., 10^5 to 10^4 years.
5. Recurrent changes, many of which appear to be cyclic, and possibly associated with fluctuations of solar output, or solar disturbance, the energy channeled toward particular parts of the Earth (especially around the magnetic poles) due to the effect of interactions between the magnetic fields of the Earth, sun, and galaxy on the travel of energetic particles, or the range of the combined tidal force of sun and moon, the periods ranging from several times 10^3 years down to less than 1 year.
6. Evolutions over 1 to 10 years in the patterns and intensity of the global atmospheric circulation, and in the temperature regime and ice extent at the surface of the Earth, which follow those great volcanic explosions that establish persistent dust veils of hemispheric or global extent in the stratosphere. It is not yet

established whether any regularity exists in the occurrences of much longer-term waves of enhanced frequency of great volcanic activity of wide, and possibly global, extent: time scales apparently in the range 10^2 to 10^4 years.

7. Changes generated by variations in the circulation and internal heat balance of the oceans, including both surface ice and the deep ocean, on time scales ranging from a few weeks up to 10^3 years.

8. Fluctuations and changes of up to several weeks' duration, associated with variations in the circulation and internal heat economy of the atmosphere or oceans. Some of these seem to be more or less regular cyclic variations, e.g., on time scales of 7, 15, or about 30 days. Some may also be related to an external "trigger", e.g., a solar flare or a particle invasion of solar or cosmic origin.

9. Variations in the transparency of the atmosphere due to cloudiness, volcanoes, smoke from forest fires, or from heavily populated and industrial areas, changes in the amount of carbon dioxide and other gases, or aerosols and water vapor.

10. Regular effects of the yearly passage of the Earth around its orbit, including all the stages of seasonal variation of radiation and the responses produced in the atmospheric circulation, the finer seasonal structure of the year, and lastly the regular effects of night and day as the Earth rotates on its axis, emanating its heat (entropy).

In conclusion it should be stressed that the Earth has always been changed by cyclic events. However, their impact on long-term climate changes was only a minor one. The decisive changes can only be explained by the first thesis advanced here: continental drift and pole wandering. Both supplement each other and are well corroborated by geomagnetic and geographic research.

C. Continental Drift

Continental drift is based on a hypothesis, generally linked with the names of Taylor,[1] Wegener,[2] and Hess,[3] of displacements, on a geological time scale and by distances of the order of thousands of kilometers, of various parts of the surface of the Earth relative to other parts. According to this hypothesis, continents now separated were once joined together as, for example, South America with Africa and North America with Europe; in general, changes in both latitude and longitude are implied.

Evidence advanced in support of the hypothesis includes similarities of shape and topographical features and of plant and animal life of land masses supposed once to have been joined. The hypothesis has been used to explain, with limited success, climatic changes on a world-wide scale. Recent evidence strongly advanced in support concerns the history of the magnetic field of the Earth as inferred from rock magnetism.

The hypothesis of continental drift has been highly controversial. Various geophysicists have held that there is no conceivable mechanism which can give rise to such an effect; in addition, various items of paleontological and paleoclimatological evidence have been held to be in conflict with, rather than in support of, the theory. The hypothesis has nevertheless now gained wide acceptance and would explain changes in climate.

II. ACCEPTED THEORIES

A. Continental Drift of the Earth and Polar Wandering

1. Geomagnetism

As early as 1686 Robert Hooke suggested polar wandering as a possible explanation of the evidence from finds of fossil ammonites and turtle shells that Europe had a

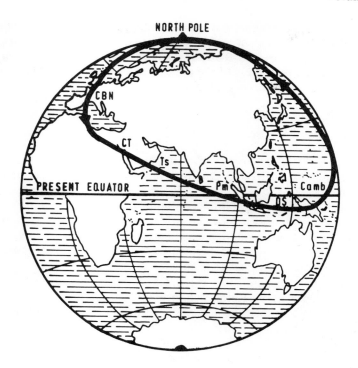

FIGURE 1.   Pole-wandering path during the Paleozoic and Mesozoic epoch indicated for the Northern Hemisphere from paleomagnetism measurable in rocks of North America and northwest Europe. The North Pole shifted from its present position to Malaya and back via Siberia and Europe to the Arctic. The South Pole shifted via South America to South Africa and thence in a great arc across Australia back to Antarctica. Such movement occurred during the following epochs: CBN = Carboniferous, CT = Cretaceous, TS = Triassic, D = Devonian, Pm = Permian, OS = Ordovician, and Camb = Cambrian.

tropical climate at some time in the geological past. It has also been alleged that in 1620 Francis Bacon, noting the correspondence in shape between South America and southern Africa, suggested that they might once have been linked together. The idea of continents drifting apart was revived by Alfred Wegener[2,31] adducing in evidence the similar rock formations on either side of the South Atlantic and the lighter, largely siliceous rock of the continent blocks floating on the denser iron- and magnesium-rich materials underlying both continents and oceans. Despite the fact that this hypothesis offered a much simpler explanation than any other advanced for the matching coastlines and geology of various continents, continental drift and polar wandering were not generally accepted until the evidence of paleomagnetism, accumulating rapidly from the 1950s onwards, was seen to support it (Figure 1).

The paleomagnetism of rocks has produced a body of evidence about the behavior of the magnetic field of the Earth, leading to the theory of the origins of this field in the motions in the fluid core of the Earth. These motions must be governed to some extent by the rotation of the whole system, so that on a long-term average the geomagnetic axis lies along the rotational axis of the Earth, though fluctuating around it with a standard deviation of some 11° of latitude.

A complication was discovered in that 50% of the measurements of remanent magnetism in the older rocks were 180° off, indicating that the magnetic field of the Earth has reversed many times, flicking over at intervals of about a million years. But, as these are precise reversals, they are no hindrance to use of the measurements as pointing in the direction of the pole at the time when the rocks were laid down. The angle of dip is preserved, and regardless of the polarity changes, makes clear in which hemisphere the place of observation was at the time the rocks were formed.

2. Continental Drift

If the rotation axis of the Earth remains fixed in attitude, as its motion around the sun proceeds through the ages, while the crust breaks into continental blocks (and smaller fragments) which part from each other and glide over the mantle and collide, we have both continental drift and the appearance of polar wandering. This is what seems to have happened, though some conflicting evidence (particularly in the earlier eras) and difficulties of interpretation continue to appear. The evidence of paleomagnetism is generally supported by the evidence of climates in each era from the distribution of corals, deserts, salt deposits, and glaciation. A difficulty, however, lies in the fact that the climatic zones each cover too broad a band of latitudes to corroborate the paleomagnetic latitudes with adequate precision.

It appears to be true that not only the paleoclimatic, but also the paleontological and geological evidence of past eras is best explained by the crustal drift positions indicated by paleomagnetic data; however each of these other lines of evidence individually yields too imprecise an interpretation to clinch the matter. Proof may ultimately elude us, but the continental drift data provide the most attractive hypothesis which is the simplest and is at least generally supported by the widest range of independent types of evidence.

It is striking that the contours of America and Euro-Africa perfectly match each other, being separated, by a deep rift, the so-called Tuscavora Rift.

The thesis which makes sense of the movements of the continental blocks is a system of convection in the upper mantle. Material rises in the midocean ridges and from there spreads to either side; the concept is known as sea-floor spreading.[3] The new oceanic crust so formed becomes magnetized and the direction of this magnetization persists in the rock concerned. Because of the variation with time of the field of the Earth this is evidence of the time when the rock solidified and hence the rate at which it has spread sideways. The rates are from 1 to 9 cm/year.

3. Polar Wandering

Polar wandering is the hypothetical movement of the axis of rotation of the Earth relative to the surface of the Earth in the course of geological time. Such movement has been advanced as a possible cause of climatic changes on this time scale. Among the evidence advanced in support of the hypothesis is that of remanent rock magnetism, on the assumption that the magnetic axis of the Earth has always been close to the axis of rotation.

The uniform dipole field which best fits the distribution of the magnetism of the Earth has "geomagnetic" (or "axis") poles at about 79°N, 70°W, and 79°S, 110°E, which define a system of "geomagnetic coordinates"; the "geomagnetic axis" joining these poles is inclined at 11° to the geographical axis. The two "magnetic" (or "dip") poles, where a freely suspended magnet is vertical, are some 1000 to 2000 km distant from the respecive geomagnetic poles. Except in higher latitudes, lines of equal geomagnetic latitude are almost parallel to lines of equal angle of dip (angle of inclination of magnet with respect to horizon), and angles between geomagnetic and geographic meridians approximate to measured angles of declination (azimuth setting of magnet with respect to geographic north). The field intensity is about 65 μT (0.65 G) near the poles and 25 μT near the equator. Decrease of intensity with height above ground proceeds approximately as the cube of distance from the center of the Earth.

Measurements of the magnetic field vector are made at a world-wide network of some 100 observatories. Continuous photographic recordings of the three independent components — horizontal force (H), angle of declination (D), vertical force (V) — are made, and the variations recorded on such magnetograms are reduced to absolute measure by regular calibration. The high standard of accuracy of about 1 nT (1 gamma) is attained in these measurements.

More than 99% of the magnetic field of the Earth is of internal origin. Observations since the 17th century have revealed the presence of slow, but cumulatively large, changes of strength and orientation of the field. Their cause is now thought to be large-scale vortexes in the conducting molten core of the Earth, slowly circulating across the main field of the Earth and so producing, by dynamo action, slowly changing regional magnetic fields. The origin of the main field itself is not yet satisfactorily explained.

Recent studies in paleomagnetism indicate some very large changes of field orientation on the geological time scale. World-wide extension of these studies has yielded substantial evidence of continental drift and pole wandering.

In conclusion it should be stated that the theory of Earth continental drift and polar wandering is well founded and is at present the only acceptable theory to explain all long-term changes of Earth climate.

B. Climatic Changes During the Tertiary Era

By the beginning of the Tertiary important features of the present global geography had come into being. The poles were in Antarctica and within an Arctic Ocean. Their shape resembled their present one though their positions were slightly different. However, the Atlantic Ocean was much narrower than the one we know and the Pacific much wider than today. Also, the Tethyan Gulf* still extended eastwards as a continuation of the Mediterranean basin, between India and Arabia on the one side and the main block of Asia on the other. Moreover, Europe was still some 20° latitude south of its present position. Africa and eastern North America were also somewhat farther south, while South America and Australia had not effectively separated from Antarctica.

In the early Tertiary the configuration of the continents was still markedly meridional, especially the western boundary of the great cluster which has been called Pan gea. It has been so for nearly 200 million years. By about 30 million years ago Australia seems to have moved far enough away from Antarctica for the great circumpolar ocean current, the West Wind Drift, to become established. It isolated the Antarctic continent from the reach of the warm ocean currents which had for so long transported heat south from low latitudes. This probably constitutes the last important piece of the present global arrangement of land and sea which brought about the Quaternary ice ages.

C. Climatic Changes During the Quaternary Era

We can now consider the sequence by which the global climatic regime passed into and out of ice ages through closer analysis of the last glaciation, since for it most evidence is available and dating errors are likely to be smallest. Variations of temperature indicators, such as oxygen isotope ratios and the abundance of various foraminifera species show many abrupt shifts of direction. Moreover, there have been some great shifts, implying changes of prevailing temperature by several degrees. They appear to have been very sharp, and were also present at about the same date in so many countries, that it cannot be doubted that the descent into the next glaciation involved

---

*   Now Mediterranean, Iranian, and Himalayan region.

several drastic events. We may also hope to gain some insight into the causation of these changes by study of their timing and by analysis of the circulation mechanisms in the atmosphere and oceans which seem to have prevailed at each stage.

The warmest period of the last interglacial was at its height around 120,000 years ago. It corresponded to the postglacial period up till now and lasted altogether some 10,000 years. According to most datings, it was ended about 115,000 years ago by an abrupt cold episode, in which there was extremely rapid growth of ice on land, to judge by study of ocean-bed oxygen isotopes. Another warm period (according to most records not quite equaling the former) followed and was ended by another abrupt cold episode about 90,000 years ago.

III. ICE AGES

A. Glaciations

1. History

The principal ice ages before the Quaternary show evidence of dependence on the continental drift and pole position results described above, i.e., the traces observable today are all in places that were in high latitudes in the eras concerned. Glaciation probably did not, however, reach equally extended limits in all the eras when the positions of the continents were favorable. One may, therefore, suppose that changes of solar output due to the evolution of the sun itself and variations associated with forces in the galaxy, or of the solar radiation reaching the surface of the Earth, played a significant part.

The following table of the principal age times has been compiled from summaries given by Harland[4] at the 21st Interuniversity Geological Congress, held at Birmingham, England, in January 1964, and by Gates[5] in a report to the U.S. National Research Council in 1974:

Circa	2250 million years ago	Archeozoic ice ages
Circa	950 million years ago	early Precambrian ice ages
Circa	750 million years ago	late Precambrian ice ages
Circa	650 million years ago	early Cambrian ice ages
Circa	450 million years ago	Ordovician ice ages
Circa	250—350 million years ago	Permo-Carboniferous ice ages
Circa	0—1 million years ago	Pleistocene ice ages

This corresponds well with the list drawn up by Tarling[5] (Table 1).

2. Glacial Centers

At various times during the last million years of geologic history, great continental ice sheets, such as now occupy most of Greenland and Antarctica, covered northern North America, most of northern Europe, and probably much of Siberia.

The European ice sheets radiated from centers located in the Scandinavian region and Scotland and extended southward into England, the Netherlands, Germany, Poland, and Russia. In Siberia, glaciers originating in the Ural Mountains, the northern uplands, and the far eastern mountains appear to have spread thinly over the surrounding plains. It is possible that several of these merged into glaciers of great extent. However, in the relatively dry environment of northern Siberia the ice sheets apparently did not attain such great thickness or extent as did those originating in the centers of Scandinavia and northern Canada. Yet they did penetrate the latest ice advance into Europe and America (Table 2).

The centers of North American glaciation were situated adjacent to Hudson Bay. From the American centers ice spread outward, but most extensively southward. At

Table 1
LIST OF SEVEN GLACIATIONS[5]

Estimated Age B.P. Million Years
Ago

Late Cenozoic	1—10
Gondwanan	225—280
Ordovician	435—500
Varangian	600—700
Sturtian	700—800
Gnejsö	900—1000
Juronian	2000—3000

one time or another it reached to a line that trends from New York westward across southern New York State to northeastern Ohio and from there nearly along the present courses of the Ohio and Missouri Rivers toward the Rocky Mountains. Adjacent to the glacier margins the ice may have been only of moderate depth, but it increased northward to thicknesses that may well have been a mile or more, sufficient at least to bury the mountains of New England. A large area in southwestern Wisconsin and adjacent parts of Illinois, and possibly of Iowa and Minnesota, was never buried by the ice. Apparently this region, known as the *Driftless Area,* was at no time completely surrounded by ice. Instead, each succeeding ice sheet passed it by on one side or the other, probably because of the blocking effect of the broad area of high ground to the south of Lake Superior.

The growth of glaciers during this "Great Ice Age" or *Pleistocene* period was not confined to the continental ice sheets. Mountain glacier systems developed or expanded beyond their normal limits on most of the high ranges of the world, though less in the tropical regions than in the higher latitudes.

3. Glacial Decline

While many of the details of glacial history are not yet known, it is generally agreed that both in Europe and North America ice sheets formed, expanded, and wasted away at four different times during the Pleistocene period. In addition, there were numerous minor advances and retreats during each of the four major cycles. By the standards of the geologic time scale, the last of the great glaciations occurred surprisingly recently, reaching a maximum perhaps 15,000 to 25,000 years ago. The final disappearance of ice from south of the Canadian border probably occurred no more than 8000 years ago. The reason for the changes in temperature and snowfall that brought the glaciers into existence and later destroyed them was probably pole wandering.

The vast quantity of water locked up in these great ice sheets was extracted from the oceans by evaporation, and it returned to them upon the disappearance of the glaciers. It is believed that the quantity was sufficient to cause a general lowering of the level of the sea by at least several scores of feet during the periods of glacier growth, and a corresponding rise in sea level during the periods of glacial shrinkage, including the present time. This would obviously have had considerable effect on the areas of the continents and the shapes of their coastal outlines. Large areas of the continental shelves, now covered by shallow seas, may well have been land during periods of great glacial advance and were inundated later. A thorough description of glaciation genesis was provided by Frakes[30].

4. Glacial Recurrence

The very first symptom of the climatic decline towards the Pleistocene ice ages can perhaps be recognized in the cool phase around 50 to 70 million years ago, when many

Table 2

THE FIVE GLACIAL AND FOUR INTERGLACIAL STAGES OF THE PLEISTOCENE AND PLIOCENE

	British Isles[a]	Alps[b]	Northern Europe[c]	Central North America[d]	Estimated age B.P. thousand years ago
1 — Last glacial	Newer drift	Würm	Weichsel	Wisconsin	120
1a — Interglacial		Riss/Würm	Eemian	Sangamon	200
2 — Glacial	Older drift Gipping Till	Riss	Saale	Illinoian	240
2a — Interglacial	Hoxnian	Mindel/Riss	Needian	Yarmouth	400
3 — Glacial	Lowestoft Till	Mindel	Elster	Kansan	480
3a — Interglacial	Cromerian	Günz/Mindel		Aftonian	550
4 — Glacial	Weybourne Crag	Günz		Nebraskan	600
4a — Interglacial	Norwich Crag	Donau/Günz			800
5 — Glacial	PreCrag	Donau		PreNebraskan?	1000

[a] Names correspond to English counties.
[b] Names correspond to German river valleys.
[c] Names correspond to European rivers.
[d] Names correspond to American states.

ancient monsters became extinct, at the end of the Mesozoic and beginning of the Tertiary. There were still some periods of great warmth to come, however, and the continuing decline of temperature levels did not set in until the middle to late Tertiary.

In the Sahara, evidence of the Ordovician ice age is more copious and better preserved than that of Permian or Pleistocene times. According to the paleomagnetic evidence the Sahara was around the South Pole at the time. Glacial formations of Ordovician age around the great Hoggar massif in the central Sahara include U-shaped valleys, floored with tillites overlaying grooved and smoothed rock surfaces. There is evidence that the sea penetrated from the northwest and that some of the grooves were made by grounded ice floes and rotating bergs. The biggest part of the ice sheet seems to have extended south of the massif.

The glaciations are the most challenging riddles of Earth history and long-term climate change. They are prominent during the salient phases of Man's development from the primate stage to the present one. It can be safely assumed that the contingencies of the ice ages furthered Man's development, born out of his three basic desires: comfort, food, energy.

B. Postglacial Times

The rise of world sea level over the last 10,000 years gives an overall view of the course of deglaciation. It is, however, a trend which must have influenced world temperature in such a way that the highest sea level — which probably occurred around 4000 years ago — should have coincided with the "end" of the period of highest temperature, which reduced the glaciers and ice sheets to their postglacial minimum. It was after 2000 to 1500 B.C. that most of the present glaciers in the Rocky Mountains south of 57°N were formed and that major readvance of those in the Alaskan Rockies first took place. At their subsequent advanced positions — probably around 500 B.C. as well as between A.D. 1650 and 1850 — the glaciers in the Alps regained an extent estimated in the Glockner region at about five times their Bronze Age minimum, after all the smaller glaciers had disappeared.[6]

1. Deglaciation

It was in the lands on either side of the North Atlantic (and to some extent, no doubt, in that ocean itself) that deglaciation was longest delayed, most of all in North America because of the huge mass of the Laurentide ice sheet. Thus, the cold regime persisted longest and extended farthest south in and near the very same ocean sector of the Northern Hemisphere. There the postglacial warming ultimately spread farthest north and from 7000 B.C. or earlier had penetrated farthest into the Arctic. The Scandinavian ice was reduced to about its present remnants by around 6000 B.C.

2. The Boreal Period (Approximately 7000 to 6000 B.C.)

At this time temperatures continued to rise. The colder seasons of the year gradually became milder, though probably with some dry and frosty winters, and the summers became generally warmer than today. Pines reached Ireland and became abundant all over England and Denmark, and in these countries the trees of the mixed oak forest made their first appearance. But the most striking feature of the vegetation of this period was the great spread of hazel throughout western Europe, including Ireland as well as Germany and as far north as central Sweden, far beyond its present limit. However, the hazel seems to have been overtaken in the next period by the advance of the great forest trees, an aspect of vegetational succession rather than of further change of climate. In northeast Finland, near 66°N, 29°E, the occurrence of sea buckthorn points to an open vegetation and a somewhat maritime climate.

3. The Atlantic Period (Approximately 6000 to 3000 B.C.)

This period spans most of the warmest postglacial times and has also variously been known as the Postglacial Climatic Optimum or the mixed-oak-forest period in Europe, corresponding to the altithermal or hypsithermal period of American terminology.

The rate of general temperature rise eased off during the first 1000 years, and in Europe the summer temperatures already may have reached their maximum by 6000 B.C. In North America, especially the northernmost parts, temperatures presumably continued to become generally higher than before throughout the period of ice melt, i.e., until after 4000 B.C. In Europe the period was probably distinguished by the mildest winters of postglacial times and everywhere north of the Alps by more moisture than in the millennia immediately before and after. The rate of rise of world sea level also slackened.

In the last 500 years or so of the period there occurred an advance of the glaciers in Europe, and probably changes in the vegetation of Europe and eastern North America (the sudden decline of the elm, ivy, and lime). Moreover, an era of migration of the primitive agricultural peoples signals a sharper oscillation towards cold climate than any for several thousand years previously. The spruce forest which had lately reached Switzerland was widely replaced by open heaths and Alpine meadows.

4. The Sub-Boreal Period (Approximately 3000 to 1000 to 500 B.C.)

After the boreal oscillations the forests regained ground in Europe, so far as Man's increasing use of the land permitted. As the open areas were recolonized by trees, the beech assumed far greater prominence than before in Denmark; oak, beech, hornbeam, and fir in Germany, with spruce also increasing in the east; in England, alders, birches, and especially oaks seem to have occupied a larger proportion of the woodlands than before. Oaks which grew in the English fenland during the warmest postglacial times, and have been found lying preserved in the peat, perhaps particularly from the Sub-Boreal times, are reported to be often of remarkably large size, e.g., with trunks reaching a height of 27.5 m before the first branch.

The climate generally regained its former warmth, and may, in some centuries, have been a little warmer than at any time since the Boreal period in Europe, but was apparently subject to recurrent fluctuations, particularly of rainfall, differentiating one century from another.

5. The Sub-Atlantic Period (Approximately 1000 to 500 B.C.)

Glacier advances, changes in the composition of the forests, and retreat of the forest from its previous northern and upper limits indicate significant cooling of world climates, its start being detectable in some places (e.g., Alaska, Chile, China) from as early as 1500 B.C. In Europe the most marked change seems to have been from 1200 to 700 B.C. By 700 to 500 B.C. prevailing temperatures must have been about 2°C lower than they had been half a millennium earlier and there was a considerable increase of wetness everywhere north of the Alps. It has been described as a period of mild winters and great windiness; cooling of the summers was presumably one of the most notable features as well.

6. The Secondary Climatic Optimum (Little Optimum)

In the European Dark Ages and early Middle Ages there was a gradual fluctuating recovery of warmth and a tendency toward drier climate in Europe over the 1000 years after 6000 B.C., particularly after 100 B.C., leading to a period of warmth and apparently of high sea level around A.D. 400. After some reversion to colder and wetter climates in the next three to four centuries, sharply renewed warming from about A.D. 800 led to an important warm epoch, which seems to have culminated around A.D.

1100 to 1300 in Europe. In these few centuries the climates in the countries concerned evidently became briefly nearly as warm as in the postglacial warmest times.

7. Conclusion

Sub speci eternitatis it seems that we are now at the end of a glacial period. Whither future periods will lead us is a crucial question.

C. The Shift of Climate in the Late Middle Ages

The general turn towards colder climates from A.D. 1200 to 1400 onwards was accompanied by shifts of the zones of cyclonic activity as the polar cap and the circumpolar vortex expanded. In the 17th century this seems to have produced a world-wide cold stage, widely known as the climatic worsening (*Klima-Verschlechterung*) of the Late Middle Ages.

In weather accounts of the time, the climatic decline is indicated by a fluctuating, but on the whole progressive, trend towards wetter summers and more frequent severe winters in temperate Europe. This trend was first apparent in Iceland and Greenland and in Russia, though the increase of storminess in and around the North Sea was also marked in the 13th century. It seemed a completely haphazard onset of harsher seasons to the inhabitants of Europe at that time but nonetheless it was clearly a time of storms and disasters more frequent than their fathers had known.

The course of the climatic deterioration over five centuries from A.D. 1200 can be traced quite well by the following facts:

1. Lowering of the tree line on the heights in central Europe and in the Rocky Mountains
2. Increasing wetness of the ground and spread of lakes and marshes in many places in Europe and Siberia, attended by swollen rivers and an increasing frequency of landslides
3. Increasing frequency of the freezing of rivers and lakes
4. Increasing spread of the Arctic sea ice into the northernmost Atlantic and around Greenland, forcing the abandonment of the old sailing routes to Greenland which had been used from about A.D. 1000 to 1300
5. Advance of the inland ice and permafrost in Greenland and of the glaciers in Iceland, Norway, and the Alps
6. Evidence of increasing severity of the windstorms and resulting sea floods and disasters by shifting sand in latitudes 50 to 60°N, in the 13th century, between 1400 and 1450 and about 1530 to 1700; the great North Sea storms of 1530, 1570, 1634, and 1694, the Hebridean storm of 1697, and the storm described by Daniel Defoe which passed across southern England in December 1703 seem to have been of a severity unmatched in the records from other times
7. Harvest failures, rising prices of wheat and bread, famines, tithes, and tax remissions
8. Incidence of disease and death among the human and animal populations
9. Abandoning of crop growing and tillage
10. Special confirmation of the trend of the climate and its effects may be seen in the records which show the decline of the English medieval vineyards

IV. FIVE ICE EPISODES

A. The Little Ice Episodes

In recent times there were some little ice episodes, the first beginning approximately

4000 B.C. and ending approximately 3000 B.C. The second lasted from 1300 to 500 B.C. The third cold spell was from 700 to 900 A.D. The fourth little ice episode lasted from 1430 to 1850, it is known as "The Little Ice Age" (Figure 2) and has been fully described in Volume I, Chapter 5. A fifth cold episode began in 1953 and has not yet left us.[7]

The ice floating in the sea off Greenland and Iceland is at present impeding shipping and fishing more than it has done in living memory — but less, so far, than it did in the 17th century, in the depths of the Little Ice Age. Then the inhabitants were sometimes completely besieged by the ice. Throughout its existence our species has contended with repeated changes of climate. Communities would adapt their hunting habits, their farming practice, or their cities to the prevailing climate, only to find after a century or two a change bringing hardship or even forcing them to abandon their lands or die.

The most recent ice age began to expire 12,000 years ago, but deceitfully. Trees and animals moved north, and up the glacier valleys, in the wake of the retreating ice as the world grew steadily warmer for about 1000 years. Then the ice had its last big fling. Quite suddenly the glaciers stopped melting away and advanced again, annihilating the impudent forests. Only after that episode did the world begin to settle down, about 10,000 years ago, to something close to present conditions.

The agriculture of the Middle East and the Indus Valley began just then. In the radically altered pattern of global climates, food supplies from hunting and gathering in those regions may have become less reliable than they were during the ice age. Thereafter the climatic conditions allowed agriculture to spread steadily westwards and northwards across Europe, from its Middle Eastern origins, without any noticeable interruptions, reaching the North Sea around 3000 B.C.

As if on the rebound, the climate in the first few thousand years after the ice age was very warm — generally warmer than anything experienced since. By 5000 B.C. Europe was two or three degrees warmer than it has been even during the warmest decades of this century. The floating Arctic ice shrank to a very small area, while the melting of the land ice was raising the sea level. Heavy rainfall in northern Europe swamped many forests and turned them into peat bogs. In that "climatic optimum" vigorous winds also brought summer rain into what is now the deep Sahara Desert.[7]

Since 4000 B.C. the story has been one of overall cooling, a slow decline that is already leading us inexorably into the next ice age. Warm periods of lessening intensity have alternated with cool periods of deepening chill. One of the cool periods (4000 to 3000 B.C.) coincided with the rise of the first cities in Mesopotamia and the founding of the first Egyptian dynasty. Glaciers are known to have advanced in Switzerland, Washington State, and Patagonia. Life was hard for people in the northern lands. But after 3000 B.C., when conditions were altogether warmer again, came the climax of the Megalithic civilization in northwest Europe, of which Stonehenge in England is a celebrated monument.

The next long cool spell ran from 1300 to 500 B.C. Glaciers that had melted completely away in the American Rockies began to reform, and glaciers everywhere advanced: Alaska, Utah, Sweden, and Patagonia give definite evidence. The widespread effects of climatic change may well have had a lot to do with the human upheavals of that time, which saw great migrations and invasions. The civilizations of the Hittites and Mycenae fell; those of the Assyrians, the Phoenicians, and the Greeks rose.

Nevertheless, as a matter of context, the empires of Ashoka in India, of the Ch'in dynasty in China, and of the Romans in Europe, North Africa, and the Middle East all grew during the warmer period that began around 500 B.C. The Romans' rivals in North Africa and the Middle East may have been weakened by generally dryer conditions, and later improvements in the north can only have helped to nurture the "bar-

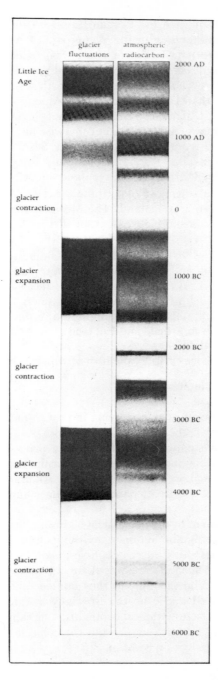

FIGURE 2. Five ice episodes. The comings and going of the glaciers are shown in the left-hand column. The next column denotes the fluctuation. The column following that denotes the atmospheric radiocarbon findings followed by the right column revealing the pertinent date. Black denotes the advance of the glaciers. An apparent rhythm of 2500 years implies that today we are still in a generally cold phase. A similar pattern is discernible in the atmosphere of the Earth, suggesting that the sun changes in the same rhythm. (From Calder, N., *The Weather Machine*, Viking Penguin Inc., New York, 1974. With permission.)

barians'' who eventually smashed the Roman Empire. Prosperity and power shifted northwards in Europe, culminating in the Viking age of A.D. 800 to 1200. Confusingly, though, for the historian of climate, there was a spell of cooling, A.D. 700 to 900, when glaciers advanced, certainly in Alaska. But the latter part of this period was very warm: the glaciers were well retracted and in North America, for example, forests grew much further north than they do today. The American midwest was warm and dry but, by sharp contrast, China and Japan were cool. Invigorated winds were sweeping them with the cold Siberian air that might otherwise have come to eastern Europe.

The Vikings had erupted from their newly warmed Scandinavian bases in the ninth century A.D., when they invaded Russia, France, and Britain. Iceland was settled in 874 and the Viking seamen discovered Greenland in 982; their voyages to North America are well known as there was no climatic reason why they should not have easily made the crossing from Greenland. The Mongols mastered much of Asia and poured westward into Russia and central Europe at the close of the warm period, in the 13th century.

The sea around Iceland was asserting itself by 1300 and the ensuing century in Europe was a period of cooling but erratic climate. Widespread famines occurred in western Europe in 1315 when heavy summer rainfall reduced the landscape to a sea of mud. The winters remained fairly mild but poor summers persisted. It was a terrible century, with the Black Death halving the population of many districts of Europe and desperate peasants in revolt in France and England. The Alaskan glaciers began creeping forward during that century.

We are still amidst a little ice episode, the turn of which is completely uncertain.

B. Impact of Prehistoric Climate on Man
1. Change of Landscape

During the last ice age, primitive men living in groups made what was probably an easy living by hunting down the big grazing animals — reindeer, bison, mammoth, etc. — on the open grassy steppe and tundra in what is now France and elsewhere in Europe south of the ice. They recorded their life and the fauna they knew in the cave wall paintings in France (at Lascaux) and Spain (at Altamira). During the retreat of the ice, probably before 12,000 B.C., bands of reindeer hunters reached as far north as Hamburg.

With the rapid development of the postglacial warming, the rivers were swollen enormously, particularly in spring and summer, by the melting ice; gravel and sand were deposited in great quantities along the river courses, lakes formed, and sometimes soon burst or quickly silted up. The landscape was changing rapidly. But the greatest change for the human population and the animals they hunted was the disappearance of the open plains, as the forest advanced north, especially in Europe. Man seems to have adapted himself more successfully than the animals. The ranges of both moved northward in Europe, Asia, and North America, but some species among the animals were lost, probably due to their reduction by Man.

2. Change of Human Race

Other great changes in the landscape were brought by the rise of sea level, proceeding over some thousands of years at the rate of 1 m a century. It seems likely that there was great loss of life, particularly among the fishing folk camped near the sea's edge and at the primitive industrial sites for making salt by evaporating sea water. It has even been suggested that in this way the end of the ice age was one of the times, rare in history, when the total population of mankind was significantly reduced.

In North America the change probably affected the early human inhabitants less drastically. The northernmost stands of forest during the ice age were never far from the ice-sheet margin; and as the ice retreated, though the forest spread north, from about 6000 B.C., the tundra belt gradually widened somewhat and, farther south, the area of grassland spread farther east from the Rockies.

England had some human inhabitants during the last ice age. Neanderthal men with their Mousterian culture (flint scrapers) survived at Creswell in Derbyshire until 50,000 years ago, when they were replaced by incoming *Homo sapiens* with an Upper Paleolithic culture.[8] Visiting groups of hunters, too, doubtless wandered north from France. Some of the Creswellian inhabitants seem to have lingered on until the arrival of the first postglacial settlers.

3. Change in Civilization

The next turning point in European prehistory in which, by a coincidence in timing, a climatic change seems to be involved is the spread of the New Stone Age (Neolithic) culture — i.e., the first agriculture — over central and northern Europe. This development in the fourth millennium B.C. more or less accompanied the transition from the warm, moist climate of the so-called Atlantic period to the more variable, and on balance somewhat drier, Sub-Boreal regime. It seems likely that it was triggered in some way by the climatic break around 3500 to 3000 B.C., which ended the former regime, produced great glacier advances, and must have severely disturbed settled human economies in regions of Europe and Asia that were marginal for either temperature or rainfall.

The earliest beginnings of agriculture and the domestication of animals can be traced much farther back, perhaps to 9000 B.C., in some localities in southwestern Asia where the ancestors of wheat and barley grew wild and where there were wild goats, sheep, cattle, and pigs.

The areas of origin of the civilizations that flourished between 4000 B.C. and the time of Christ are, except in the special case of Egypt and the lower Nile, areas of rather intricate geography, within which the sites then occupied range today from the natural domain of woodland to steppe or desert. Details of the changes that have taken place in the environment there could hardly be shown on a world map and have scarcely been probed so closely that valid regional maps of the former vegetation could yet be drawn. In at least three regions, however, there are unmistakable indications of formerly more extensive water supply and vegetation:

1. The western and central Sahara, where rock drawings of a wide range of animals, and finds of skeletons, including elephants, show that there was enough surface water for animals to pass across the terrain that is now desert
2. The Rajasthan (Thar) desert areas of India and Pakistan, where rivers have disappeared since the Indus Valley civilization and the cities of Mohenjodaro and Harappa were at their height before 2000 B.C.; also the other desert areas in southern Asia crossed by Alexander's army on expeditions in Iran and on the march to the Indus between 330 and 323 B.C.
3. Crossing of Sinkiang, particularly the Tarim basin in central Asia, by the ancient Silk Route which was used by trading caravans between China and the West in Roman times, when there was a chain of cities and settlements there and remnants of forest; today it is largely a sandy desert

Other places, where conquest of the area by desert conditions within the last 2000 years suggests a continued trend towards natural desiccation, include Palmyra and Petra on the fringe of Palestine and the Syrian desert, and parts of the Anatolian plateau in central Turkey. Certainly, the extent of the oases has been declining over the last 6000 years.[9]

C. Climatic Changes in Historical Times

1. Historical Features

Interest in the subject of climatic change was aroused during this century by the warming trend which persisted from the late 1890s through to the 1940s; although this trend was recognized in discussions of the Royal Meteorological Society as early as 1911, widespread public discussion of the implications of continued warming only occurred after the trend had already reversed, in the late 1940s and early 1950s. In recent years, there has been some public concern that the cooling trend which followed the warming of the first half of this century might now persist to produce severe problems. The lesson of the 1950s, when there was some speculation that by the end of this century the Arctic might be ice-free but in fact the climatic trend had already reversed, warns of the dangers in such naive extrapolation, and of the necessity to examine the patterns of climatic change over as long an historical period as possible in order to obtain a more reliable guide to the likely limits of climatic change in the years up to the end of this century and beyond.

The first and most important lesson of the historical record is that there is no such thing as a climatic "normal" in the sense that the word was used 50 or more years ago, implying that if a long enough series of observations could be amassed it would produce an average value of each element describing the climate, an average to which the climate at the observing site would always tend to return. Far from being "average weather", climate is always changing on a variety of different time scales; the problem facing us is to determine the extent of these changes and to unravel the different time scales of change as far as possible from the historical record, in the hope that this will provide information about the underlying mechanisms of climatic change which interact to produce the complex pattern. Only with some understanding of which mechanisms dominate the changes occurring at present can there be any hope of reliable prediction of climatic change.

2. Climate Drift

There is evidence that some part of the pattern of climatic change drifts westward around the globe at an average rate of 0.6° of longitude per year, so that a warming recorded in the Far East in the 16th century A.D. reached Europe in the 18th century. This may explain the nonsynchroneity of the European and Chinese records, and also offers an intriguing potential insight into the global processes of climate, although this westward "wave" accounts for less than half of the variance of mean winter temperatures (taken as decadal averages) for the past 1000 years and seems distinct from global trends of warming and cooling.

Historical records from the Southern Hemisphere are almost nonexistent, and this has been a particularly worrying aspect of the attempt to describe global changes in climate from the historical record. Recently, however, a major effort to investigate the patterns of climatic change in Australia and New Guinea has been carried out by groups using the techniques of pollen and isotope analysis, changes in lake levels, glaciers, and tree lines, and geological evidence. These data cover a far longer time scale than the historical records available for some parts of the Northern Hemisphere, but they show that major changes in climate, at least, occur with near synchroneity in the two hemispheres. However, this may not be the complete story since there is some evidence from a study of temperature changes near New Zealand and elsewhere during this century that much of the Southern Ocean zone cooled between 1900 and 1945, and has since experienced a warming trend — the opposite of the pattern deduced from data from the zones north of 30°S. Again, this pattern, if real, must indicate some fundamental feature of the global situation; however, the lesson seems to be that while historical records from northwest Europe probably do provide a good guide to the broad features of recent climatic change in the Northern Hemisphere, we cannot with confidence claim that the smaller and shorter-term changes in particular can be interpreted as global trends. With that in mind, we must make the best use of the data we do have, which require extensive interpretation if the historical record of climate is to be extended back before the 1650s to the period when the barometer and thermometer were not available and no direct, quantitative records of temperature and pressure could be made.

3. Four Climatic Epochs

Four climatic epochs since the end of the most recent phase of glaciation some 10,000 years B.P. are of particular interest and merit further study. These are

1. The postglacial climatic optimum or warmest times — the old term "climatic optimum" is now out of favor — culminating between about 7000 and 5000 years B.P.
2. The colder climatic epoch of the Iron Age, culminating between about 2900 years B.P. and 2300 years B.P.
3. The secondary "climatic optimum" of the early Middle Ages, roughly from 1000 to 800 years B.P.
4. The Little Ice Age, a cold period particularly marked between about 550 years B.P. and 125 years B.P., peaking in severity in Britain in the late 17th century.

These outstanding phases of climate are of very different durations, and of course much less information is available about the older phases; it is unlikely that the complex of causes was the same for each, but they represent the extremes of recent conditions. In some cases extraneous factors — such as volcanic dust in the atmosphere — may have been particularly important.

a. The Postglacial Warmest Times

Between about 5000 and 3000 B.C. the sea level was rising rapidly as the last remnants of the former great ice sheets melted after the Ice Age, and around 2000 B.C. it may have been higher than today by about 3 m. This implies roughly 10^{15} m^3 less glacial ice than today on the mountains of that time, but with a distribution of the remaining ice not unlike that of the present climatic regime. The minimum extent of the ice cover on land was reached in the very last stages of the main warm period. Although the rise in sea level from the end of the previous full Ice Age was primarily due to the reduction of the extent of ice on land, care is needed in interpreting the records because of isostatic effects associated with the loading and unloading of ice on the land which altered the geography near the ice sheets. During this optimum, the extent of sea ice on the Arctic Ocean was also reduced, with open water extending beyond the channels of the Canadian Archipelago. A variety of data, including fossil marine fauna from Spitsbergen and evidence of a wetter regime than today in the Sahara and the deserts of the Near East, all help to fill in the overall picture, with vegetation belts displaced northward and temperatures in Europe in summer averaging 2 to 3°C above those of today. For the period up to 4000 B.C. in particular, the data are consistent with a displacement of the subpolar depressions and the axis of the main anticyclone belt northward, perhaps placing the high pressure belt as far north as 40 to 45°N.

With the high pressure belt well north of the Mediterranean, Africa and the Near East would have experienced a significant shift of the trade wind and equatorial zones, with more widespread monsoon rains in the summer, and winter rainfall as well in the Mediterranean region. This seems to be consistent with other Northern Hemisphere data, and during the climatic optimum period Hawaii also experienced increasing rainfall, presumably associated with a northward shift of the trade wind zone; in addition, Southern Hemisphere data show that this was a global climax of warmth. The southern data come primarily from studies of the extent and distribution of forest species in southernmost South America and New Zealand and from the recent work in Australia and New Guinea. The Antarctic also was warmer at about this time, with temperatures 2 to 3°C higher than now in the Antarctic continent, Tierra del Fuego, and the Himalayan Mountains.

b. Iron Age Cold Epoch

Although the period between 900 and 300 B.C. or somewhat later was a cold epoch, the most striking feature from the European record is the associated increase in wetness, with evidence of a widespread regrowth of bogs after a much drier period. This change, producing a so-called "recurrence surface", is a conspicuous feature of peat bog sections from all parts of northern Europe, including Ireland, Germany, and Scandinavia; adaptation and abandonment of ancient tracks across the increasingly marshy lowlands in England and elsewhere provides confirmatory evidence. The forests of Russia spread further south at this time than during the preceding warm epoch, and a change in the dominant species indicates some lowering of the summer temperature.

Further south, in the Mediterranean region and in North Africa, the climate during the Iron Age cold epoch seems to have become drier, but not so dry as it is today. Roman records are of value here; agricultural writers, such as Saserna, noted around 100 B.C. that the vine and olive were spreading north in Italy to regions where the weather had previously been too severe, suggesting that the centuries before 100 B.C. had indeed been cooler in the Mediterranean. Southern Hemisphere data also indicate a cooler period roughly around 500 to 300 B.C.

c. Secondary Optimum

For the climatic optimum which occurred around 1000 to 1200 A.D. and later epochs, more comprehensive data are available and historical records assume greater importance.

The secondary optimum shows many of the same features as the postglacial optimum, but to a lesser degree and for a shorter duration. Melting of the Arctic pack ice and restriction of the southward spread of drift ice played an important part in the founding and development of Norse colonies across the northern North Atlantic to America; in western and central Europe vineyards extended 3 to 5° further north in latitude and 100 to 200 m higher above sea level (for a discussion of the vine as a climatic indicator, see present day); indirect evidence suggests that these temperatures had in fact been reached by the 1600s.

Forests at high levels in central Europe declined catastrophically after 1500, and eyewitness reports from Scotland describe widespread dying off of woods in exposed localities. But although winters were very severe, it seems that in Europe around 50°N there were times when the prevailing summer temperatures were mostly around their present level, and even slightly above in the 1700s. Some evidence suggests a greater year-to-year variability of summer and perhaps of winter temperatures than in the present century. Severe winters also affected the Mediterranean, and glaciers advanced generally in Europe, Asia Minor, and North America, while snow was recorded lying for months on the high mountains of Ethiopia where it is now unknown. The Caspian Sea rose during the Little Ice Age and maintained a high level until 1800, but the White Nile, fed from the rains of the equatorial belt, was low and these rains seem to have been weakened or shifted south. The overall picture is one in which there was an equatorward shift of the prevailing depression tracks in the Northern Hemisphere, with more prominent polar anticyclones.

This pattern is not fully reflected in the Southern Hemisphere, however, which seems to have partly escaped the cold epoch until the 19th century. Between 1760 and 1830 the fringe of Antarctic sea ice seems to have been a little south of its present position, with the southern temperate rainbelt also displaced southward — that is, *away* from the equator. After about 1830, until 1900, the rain zone and depression tracks moved north with advances of the glaciers in the Andes and south Georgia at about the time that the Northern Hemisphere was recovering from the Little Ice Age.

While the first three epochs seem to have been associated with global shifts of climatic zones towards or away from the equator roughly synchronously in both hemispheres, the Little Ice Age data suggest a different pattern, with an equatorward shift in the Northern Hemisphere accompanying a synchronous poleward shift in the Southern Hemisphere, followed by return movements in both hemispheres, though the *temperature* trends over most of the two hemispheres may have been more nearly parallel. The amplitude of surface temperature fluctuations has been greatest at high latitudes so that the meridional gradient at the surface, at least between 40° and 6°N, must have been markedly less in the warm epochs than in the cold ones; but temperatures on mountains even in low latitudes were 2 to 3°C higher than today in warm epochs and 1 to 2°C cooler than today in cool epochs. Possibly the meridional equator-pole gradient of upper air temperatures was actually less during the cold epochs (References 13 and 21). This evidence, together with estimates of the limits of tree lines, implies mean summer temperatures a little more than 1°C higher than those now considered normal.

Archaeological and other paleological evidence from North America suggests that the secondary optimum lasted until about 1300 A.D. before there was a change to cooler, wetter conditions; to the south there is evidence of a wetter period contemporary with the little optimum, but perhaps longer lasting, in Central America, Cambodia, the Mediterranean, and the Near East. In Antarctica, there is evidence that the penguin rookery at Cape Hallett was first colonized just before this period, presumably during a phase of improving climate, and that it has since remained occupied permanently.

d. The Little Ice Age

Although the climate of the past 1000 years was at its most severe in Britain during the second half of the 17th century, the Little Ice Age is generally taken as the longer period, from 1430 to 1850, during which there was a predominance of harsher conditions than those of this century but also some periods of relatively equable climate. During this epoch the Arctic pack ice expanded considerably, with important effects for Greenland and Iceland, and by 1780 to 1820 temperatures across the North Atlantic everywhere above 50°N seem to have been 1 to 3°C less than those of the surface temperatures at high latitudes particularly unrepresentative at such times because of frequent strong inversions.

There is at least a hint here that the climatic changes of the Little Ice Age had a somewhat different pattern of causes from those of the three other epochs which, with the Little Ice Age, represent the greatest departures from present-day conditions since the end of the latest full glaciation. This is particularly interesting since the Little Ice Age is the most recent of these epochs, and it is even possible that the warming of the late 1800s and early 1900s represents a temporary fluctuation from Little Ice Age conditions, to which we may shortly return.

e. Conclusions

It is clear from the pattern of past climatic changes that naive extrapolation of any trend is no reliable guide to future prospects. In the early 1950s, there was concern lest the warming trend of the first half of this century should persist, although soon it became clear that the increase had stopped. A short-term guide to events which may be of long-term importance comes from studies of changes in marine populations and the regions favored by different species. Over the 24-year period 1948 to 1972, the total numbers of copepods and the zooplankton biomass in the North Sea have declined, the duration of the zooplankton season has become progressively shorter, and the spring bloom of phytoplankton progressively later. Southerly shifts in the distribution of fish have occurred, with a significant movement from 1965 to 1967 which seems "likely to have had its origin in climatic events". Since 1940, an average cooling of the Northern Hemisphere by 0.4°C has reduced the length of the growing season in England by 10 days and has undoubtedly played a part in recent agricultural troubles in other parts of the world. This stresses the question of whether the future will bring back the favorable climatic conditions that prevailed in part of this century, or will remain closer to the cooler conditions that characterized most of the past six centuries or more. On the face of it, the recent warm peak was an unusual event. Bryson[22] has summarized "the lessons of climatic history" as:

1. *Climate is not fixed.*
2. *Climate tends to change rapidly rather than gradually,* with changes from a glacial to nonglacial climate, and vice versa, perhaps taking less than a century, although full response and adjustment of the environment may take much longer; smaller but still significant changes occur over a few decades.
3. *Cultural changes usually accompany climatic changes,* as shown so clearly by the history of the Norse colonies.
4. *What we think of as the normal climate at present is not normal in the longer perspective of recent centuries.*
5. *When the high latitudes cool, the monsoons tend to fail* — and the high latitudes have now been cooling for some 25 years.
6. *Cool periods of Earth history are periods of greater than normal climatic instability,* associated with weaker circulation, bringing extremes of drought, flood, and temperature (including even occasional extreme summer heatwaves) to further disturb agriculture, already handicapped by a short growing season.

The future of climatic changes thus remains ambiguous.[23]

f. Three Special Climate Changes in Historical Times

Glaciation — A number of severe winters had been reported in ancient Rome with mentions of the Tiber being frozen in 398, 396, 271, and 177 B.C. On the other hand, the Roman agricultural writers Saserna (father and son) wrote that in the last century B.C. cultivation of the olive and the vine were spreading farther north in Italy where in the previous century winters had been too cold for the plants to survive. The next several centuries seem to have been an easier time, the reports of occasional "severe" winters then being associated with flooding rather than ice on the Tiber. Some indication of a rather long immunity from harsh winters in Europe is provided by the history of the bridge with many stone piers which Apollodorus of Damascus built in A.D. 106 for the Emperor Trajan across the Danube at the Iron Gates into what is now Rumania. This bridge stood for almost 170 years without being carried away by ice and was reputedly destroyed by the Dacian tribes when the Romans withdrew from the area. The glaciers in the Alps seem to have been in retreat from about 300 B.C. to A.D. 400,[10] and that they were for some time not so extensive as now is indicated by Roman gold mines high up in the Alps in the Sonnblick area (Austria), some of which are probably still under the ice since others have only lately come to light as the ice receded. Traffic over the Alpine passes continued even in winter time.

Inundation — There was, perhaps, a final climax about 120 to 114 B.C. which altered the coasts of Jutland and northwest Germany in a great sea flood, the "Cymbrian flood", which set off a migration of the Cymbrian and Teutonic peoples who had been living in those areas. Lesser episodes of the same kind appear to date from about 2200 B.C. and A.D. 500. The general impression of a rather colder climatic period in or about the third quarter of the first millennium A.D. receives some elucidation in that the summer temperatures point to separate climaxes of cool summer climate: (1) between about 500 and 700; (2) in the 9th century; and (3) in the 11th century A.D. The balance of the winters seems to have been coldest in or around the 3rd to 4th and the 8th centuries, and again between about 850 and 1000 A.D.

Droughts — The Caspian Sea was at its lowest level in the last 2000 years at about A.D. 300 and 800, and these two epochs were noted by Huntington[11] as times of abandonment of settlements all along the route through Chinese Turkestan that had connected China with the West. The level of the Caspian Sea remained generally low until the 12th century. Droughts come rather notably to prominence after A.D. 350 among the scattered reports available from Britain and central Europe until about 484. They were immediately succeeded by a great predominance of flood reports in the next century or more, particularly in the 550s and between 575 and 590, which was a very wet period over much of Europe. The warm summer periods around A.D. 300 to 500 and 700 to 800, and again in the 900s, were accompanied by general dryness and cold winters. The warmth of the high Middle Ages in Europe between A.D. 1000 and 1300 (most of all in the late 12th and 13th centuries), and the more moderate warmth around the time of the 1st century A.D., were accompanied by mild wet winters; yet both these periods may have been generally moist in the Mediterranean with an incidence of thunderstorm rains in summer quite unlike the present day. Indications of this are to be found in the diary of Claudius Ptolemaeus of Alexandria about A.D. 120 and in the existence of Roman bridges across Arabian rivers and wadis that are now dry. An even more remarkable case is the five-arched medieval bridge at Palermo, Sicily, built in A.D. 1113, to span a river which was then navigable but hardly exists today.

V. MOOTED TERRESTRIAL CHANGES

A. Sea Level Changes

Evidence points to a somewhat lower general sea level for some centuries after 500 B.C., a higher level during (and for some time after) the warmest periods in late Roman times and the Middle Ages, and a lower level again in the 17th and 19th centuries A.D.

Evidence of higher general sea level in the Early Middle Ages includes:

1. The watery state of the English Fenland, among the rush-and-willow-girt islands on which the Anglo Saxon hero Hereward the Wake was able to hold out for 10 years (1066 to 1076) against William the Conqueror.
2. The existence about A.D. 1000 (and reputedly also around A.D. 400) of an East Anglian inlet like a Danish fjord, in which the sea reached inland to Norwich.
3. The engulfing of a huge area of the Netherlands to form the Zuyder Zee through the agency of the storms in 1250 to 1251 and 1287.
4. The similar loss to the North Sea of great areas off the Belgian, Danish, and German coasts (Bruges was once a harbor).
5. The use by ships of sea channels which have since dwindled or entirely disappeared — Limfjord in northern Jutland as a passage from the North Sea to the Baltic avoiding the Skagerrak; another sheltered channel reputed to have been used by Danish Viking ships on the south side of the Moray Firth in Scotland, joining Lossiemouth to Burghead Bay; and the channel, known as the Minster Fleet, which separated the island of Thanet from the mainland of England.

The Caspian Sea, after its low levels around A.D. 300 and 800, continued lower than now until about A.D. 1200 but then rose sharply over the next two to three centuries, being in the 1400s up to 8-m higher than now. It is clear that a notably moister regime spread across central Asia, but this has to be regarded as part of the climate shift of the late Middle Ages towards the oncoming Little Ice Age. The earlier medieval warm epoch had been a relatively dry time in those regions.

Changes of climate can be anticipated from fundamental changes in the direction of ocean currents, especially the warm Gulf Stream. This problem has been discussed by Brooks,[12] however no such change has yet been described. Summarily, we can say that the critical review of the role of sea or ocean current changes can hardly ascribe to them any relevant effect on long-term climatic changes.

B. Earth Elevations
1. Folding

There have been long periods in which the crust of the Earth was at rest, while the denuding agencies gradually lowered the surface almost to a uniform plain and the waves of the sea bit deeply into the continents. Alternating with these have been relatively short periods of intense disturbance during which the surface of the Earth was thrown into great folds and ridges, when the mountain ranges which form the articulated skeletons of the continents were brought into being. The greatest periods of mountain formation occur in close relation with the greatest periods of glaciation; thus the Alpine period of folding in the Tertiary preceded the Quaternary Ice Age; the Hercynian* folding in the Upper Carboniferous preceded the Upper Carboniferous glaciation. It is known that there was a period of great disturbance and mountain building preceding the Cambrian period, and another, lower in the Proterozoic, probably preceded the first of the four great glaciations, the deposits of which, in Australia at least, rest on great outflows of lava. The minor cold period of the Silurian was also associated with a period of folding and mountain formation, the "Caledonian", which, however, did not reach the intensity of the Hercynian and Alpine foldings. Thus we can represent the variations of mountain building activity and of climate during geological time as a series of waves in which the long troughs represent the periods of stability and genial climate, the sharp crests the periods of mountain building and climatic stress.

The last two ice ages at least were not synchronous with the maximum of mountain formation, but followed them after some millions of years. This lag has been attributed to three causes:

1. After a long quiescent warm period the whole mass of the oceans was warm and had to be cooled down before general glaciation could begin. This process would occupy thousands of years and smooth out climatic fluctuations within an ice age, but could not cause a lag of millions of years.
2. The second cause is that mountain ranges are first elevated as smooth domes, which are worn into irregular contours by the ordinary processes of erosion. The lightening of load caused by the removal of this eroded material causes further isostatic elevation and a greater effective height.
3. The steady warming of the crust of the Earth by radioactivity is much greater than the normal escape of Earth heat at the surface, so that the crust becomes continually hotter and more plastic. This allows folding, mountain building, and volcanic outbreaks, in which the accumulated Earth heat is liberated.

---

*   Harz Mountains (Germany).

2. Volcano Activity

Volcanic dust is known to have spread in the stratosphere as a veil covering more than half the surface area of the globe in some instances and to have persisted in observable quantities for up to 3 years. The latitude zones which are sooner or later affected probably depend greatly on the latitude of dust injection. Such dust veils are associated with certain atmospheric optical effects (e.g., Bishop's Ring*). It is also probable that significant effects on atmospheric circulation and world weather are caused by the scattering of solar radiation by widespread and persistent veils.

According to Lamb's[13] estimate of the magnitude of eruptions dating later than 1600 (estimate based on evidence of dust veils and/or of quantities of solid material ejected), meteorological effects were probably significant in at least the following cases, listed in approximate order of magnitude: 1783, Skaftárjökull or Laki (Iceland), Eldeyjar (Iceland), and Asama (Japan), effects lasting until 1785; 1815, Tambora; 1883, Krakatoa, effects lasting until 1885 to 1886; 1680, Krakatoa and Tonkoko; 1831, a group of major eruptions including Pichincha, Mediterranean submarine eruption, Babuyan, Etna, and Vesuvius; 1821 to 1824, a group of major eruptions including Kluchevskaya Sopka in 1821, Eyafjallajökull (Iceland) from 1821, Vesuvius in 1882, Galunggung in 1822, and Lanzarote (Canary Islands) in 1824; 1902 to 1904, a group of major eruptions including St. Vincent (West Indies) from 1902 to 1903, Mont Pelée in 1902, Santa Maria (Guatemala) from 1902 onwards, and Colima (Mexico) in 1903.

In addition to these cases, thick volcanic ash layers in various parts of the world supply evidence of former volcanic activity of a magnitude likely to have had significant climatic effects. In some cases they can be approximately dated. It appears that volcanic activity was particularly frequent from A.D. 1500 to 1900, around 500 to 0 B.C., around 3000 B.C., and around 7500 B.C. Both hemispheres were apparently affected by the three earlier waves of activity, but the A.D. 1500 to 1900 period seems to have affected mainly the climates of the Northern Hemisphere and the equatorial zone.

1627 BC

Marinatos[14] has shown that the eruption of a volcano on Santorini in 1500 B.C. destroyed the whole Minoan culture of the Dodecanese Islands. His excavations at Thera divulged a superb culture covered under 50 to 60 m of volcanic lava. This eruption of Krakatoan extent reached as far as Crete and affected the climate of the Mediterranean basin.

3. Surface Cycles

The surface of the Earth has passed through a series of cycles, each cycle consisting of a relatively short stage of intense mountain building, in which the rocks were thrown into great folds and ridges and the average elevation of the land above the sea became very great. This was followed by a long stage of quiet conditions, in which the forces of denudation lowered the level of the land, rapidly at first, and then more and more slowly. We are at present living shortly after one of the periods of mountain building and elevation, and the average level of the land is consequently high, though not as high as it was during the Quaternary period. At the close of one of the long quiet periods, the average land level must have been very much less than it is now, and was probably only a few hundred meters. At such times, conglomerates and coarse marine sandstones are almost entirely absent, and the bulk of the sedimentary rocks is composed of limestones and very fine clays or shales.

In conclusion it can be stated that Earth elevation and folding had only a minor contributing effect on long-term changes of climate.

---

*   Volcanic dust corona.

C. Earth Heat

1. History

The connection of Earth heat to long-term climate changes would appear negligible, since it is no more than 0.1 to 0.2°C above that of the Earth crust.

Within its protecting cloud canopy, the surface, oceans and continents alike, was warm from equator to poles, but the land surfaces cooled more quickly than the heat conserving oceans, and in due course, while the warm oceans were still supplying enough moisture to maintain the cloud canopy intact, the land surfaces began to freeze, and ice sheets developed. Apart from some local glaciations in the centers of the larger continents this stage was first reached on a planetary scale in the Permo-Carboniferous period; this glaciation coincided more or less with the present subtropical high pressure belts, and the reason is stated to be that "cold anticyclonic winds" cooled the land most rapidly in those belts. The cooling of the oceans continued and with decreasing evaporation a stage was reached in which these high pressure areas, today possessing the clearest skies of the world, ceased to be mantled in clouds — the sun broke through and deglaciation commenced.

Now followed a period of dual control, solar energy prevailing near the equator, Earth heat towards the poles. In spite of fluctuations, the latter gradually diminished, and just before the Quaternary glaciation the polar oceans became cold for the first time. Then the second planetary glaciation occurred, centered in the cold temperate belts of greatest precipitation, at this time the only regions which were permanently overcast. The cooling of the oceans continued, and evaporation ceased to supply enough moisture for even this limited cloud belt, the sun shone over the whole world, deglaciation again commenced and is still continuing.

2. Impact of Earth Heat on Climate Change

The theory is interesting, but there are some insuperable difficulties. With warm oceans and an unbroken cloud canopy, the land surfaces, unless at a great altitude, would not be likely to freeze; the conditions are most nearly realized at present in the equatorial rain belt, in which the land is maintained at the same temperature as the neighboring oceans. "Cold anticyclonic winds" presuppose cooling by radiation; even if under world-wide isothermal conditions the pressure distribution could remain unaltered, which is highly improbable, we must suppose either that the anticyclone would break down the cloud canopy, in which case the tropical sun would certainly prevent glaciation, or that the clouds would remain in spite of the anticyclone, in which case the descending air would not be cold.

Summarizing, the theory of Earth heat changes can hardly explain long-term climate changes.

3. Warming of Polar Ice Masses

As seen on a geological time scale of tens of millions of years, we now find our planet in the "unusual" condition of having some permanent ice and snow in both polar regions. For roughly 90% of the time in the past 500 million years the poles have been virtually ice-free, at least in summer. However, on a time scale of hundreds of thousands of years we are in an unusually warm period, and there are no longer massive ice sheets covering northern North America and northern Europe, as there were 18,000 or 20,000 years ago — the vestiges of which were apparently still in retreat only 8000 years ago. Thus, the degree of polar glaciation, or the extent of ice at high latitudes that survives the summer, has varied greatly in the past.

These changes of the quasi-permanent ice masses must have been the result of natural processes governing the heat balance of the Earth, processes connected with the ocean circulation, the orbit of the Earth, the positions of the continents, the sun's output, and so forth. Now we are about to change that heat balance again.

Of the four regimes of ice on Earth, the two that are probably most important to consider are the floating sea ice, and the great ice sheets. The floating sea ice in the Antarctic appears and nearly disappears each year, while in the Arctic Ocean there is always a substantial area of sea ice the year round. The contrast between the two polar regions can be illustrated by the fact that the area of pack ice frozen each winter (and melted each summer) around the Antarctic Continent is larger than the area of the entire Arctic Ocean.

Referring to the Arctic Ocean specifically, the major question is whether a large warming can result in removing the pack ice completely, and whether such a complete removal will mean that it will remain open and not freeze over again in winter. There are several reasons for arguing that it probably would tend to remain open once the ice pack had been melted, barring a major change in sea level. For one thing, the Arctic Ocean would present a dark surface in summer compared to the highly reflecting ice pack that exists now, so, even with low clouds covering the area, a great deal more solar energy would be absorbed by the system. Another rather compelling reason for thinking that the Arctic Ocean would be harder to freeze over once the ice pack had been removed is that there is now a layer of relatively low salinity water floating under the ice pack to a depth of 10 to 30 m, and since this relatively fresh water has a lower density than the normal salt water of the ocean, it produces a stable layer that inhibits mixing and exchange of heat between the surface layers and the warmer waters below. With the ice pack removed, wave action and surface currents would almost surely eliminate this thin stable upper layer. For both of these reasons, it is expected that if and when the Arctic Ocean ice pack is removed the open, freely mixing ocean will not freeze over again until another very major cooling of the Northern Hemisphere occurs.

4. Conclusions

Evidence which could show whether a major Antarctic ice cap surge actually caused any Pleistocene glaciation has been discussed by Hollin[29] and includes sharp excursions of ^{18}O levels in deep sea cores, ice cores, and other records. Flohn[27] refers to the sharp changes around 95,000 to 90,000 years B.P. in the Mexican Gulf cores, Orgnac stalagmite, and Camp Century core as evidence of "instant" glaciation following an Antarctic surge. Perhaps most strongly supporting the surge ice age hypothesis would be evidence showing a sudden rise of sea level, indicating sloughing of considerable Antarctic ice into the sea, immediately followed by slower eustatic fall corresponding to northern glaciation. No such eustatic pattern has been identified yet, although a possibility is indicated by the rise to the 120,000 years B.P. high sea level peak. The rate of this rise has yet to be determined, and hence the surge glaciation hypothesis cannot with finality be ruled out for initiation of the last Ice Age period.

VI. MOOTED COSMIC CHANGES

A. Astronomical Influence
1. Changes of Earth Orbit

The three very long cyclic variations in the orbital arrangements of the Earth first calculated by Milankovitch[15] must be supposed to have effects upon the sun radiation regime, and through that upon wind circulation and other climatic conditions, as inescapable as those that accompany night and day, winter and summer. These are

1. The varying ellipticity of the orbit — period circa 96,000 years
2. The varying obliquity of the ecliptic due to tilt of the rotation axis of the Earth — period circa 40,000 years
3. The precession of the solstices and equinoxes — period circa 21,000 years

The effects of these variations on the input of solar radiation at various latitudes in the summer and winter, respectively, correspond to superposed sine curves of differing wave length, amplitude and phase, producing an at first-sight irregular sequence of fluctuations of varying amplitude. Because the time scale of these variations is so long, their effects upon the atmosphere cannot be demonstrated with certainty nor in detail to compare with those that accompany night and day or the seasonal round of the year. Nevertheless, it is widely accepted that they have to do with the sequence of ice ages and interglacial periods and with the warmer and colder stages within each.

The relationship of the dates of the Milankovitch[15] radiation fluctuations to the full terrestrial sequence of ice ages and interglacial warm periods in the Quaternary era cannot yet be adequately tested. Unfortunately, independent methods of dating the ground evidence from more than 50,000 years ago still depend too much on extrapolating sedimentation rates in the sea bed and lake bed deposits examined or growth rates in bogs. Nevertheless, the assumption of a terrestrial response to long astronomical cycles affecting the radiation supply (with whatever lag is imposed by the inertia of large accumulations of ice) rests on the firmest possible theoretical justification.

2. Tidal Forces of the Planets

The varying tidal pull of the differential attraction of the moon on the Earth and its fluid envelope, exerted by the planets, can control the disturbances on the face of the sun in the form of sunspots. Alternatively, the various observed disturbances of the sun and variations of its energy output may be entirely due to powerful thermal and thermochemical processes inside the sun or to some interplay between these and the varying tidal force of the planets.

Jupiter and Venus, at their mean distances, contribute most to this tide raising force of the sun, respectively, about 34.6 and 33.0% of the total. The contribution of the Earth at its mean distance is approximately 15.5%. Together these three planets account for 83% of the total. Conjunctions of Jupiter and Venus occur at 237-day (0.65 year) intervals and oppositions at half this time. Conjunctions of the Earth, Jupiter, and Venus occur at intervals of 24 years; alignments, with either Jupiter or Venus in opposition, occur every 12 years. These alignments and conjunctions are reinforced by alignments with Mercury and Saturn at intervals of 60 and 120 years, respectively, and by alignment with Uranus at intervals of 84 and 168 years, respectively. It is evident that all these planets return to similar positions, with the major ones being in conjunction, at intervals of 1680 years. Alignments, with one of the main contributing planets in opposition, occur at the half-period of 840 years.

There is, thus, a *prima facie* case for accepting the existence of some long cycles in weather phenomena which may be of about the right order of length to agree with the planet-induced sun-tide cycles noted. The amplitudes of the climatic fluctuations mentioned on time scales of about 1700 to 2000 and 3500 years appear to be great in terms of rainfall in the heart of Asia and of summer and winter temperatures in Europe and central North America. These fluctuations presumably operate through changes in the prevailing latitude, strength and wavelength of the mainstream of the upper westerlies, and the associated thermal gradient. Unfortunately, there is little prospect of adequate precision in dating the evidence to establish how nearly the long periods in weather and external phenomena agree.

3. Tidal Forces of Sun and Moon

Sun and moon both produce semidiurnal tides of 24 hr, 50 min in the atmosphere as in the oceans. The double wave is attributable to the gravitational attraction of either body, acting upon the atmosphere and ocean on the near side of the Earth and acting upon the Earth itself to pull it away from its fluid envelope on the far side. The lunar tide in the atmosphere is, however, so minute as to be of no practical importance. The average tide-raising force of the sun on the Earth is about half that of the moon, and in the ocean the sun raises a tide about half the height of the lunar tide. When the tidal pulls of sun and moon reinforce each other at about full moon and new moon, the total range of the spring tide is actually about three times that of the neap tides at the lunar quarters.

The moon's orbit is tilted at 5°9' to the ecliptic, and for this reason the extreme declination which it attains on its way around the Earth varies. The maximum declination in the present epoch is about 28°40'N and S: the moon ranges this widely at intervals of 18.6 years. Midway through each 18.6-year period the declination of the moon ranges only between about 18°20'N and 18°20'S. Recent maxima of lunar declination were in 1876, 1894, 1913, 1931, 1950, and 1968. The tidal force of the moon acting upon high latitudes is greatest when its declination is greatest, and particularly when its perigee position in its orbit coincides with maximum declination.

The combined tide-raising force of sun and moon must produce the greatest tidal range in the oceans when Earth, moon, and sun are in closest alignment and when the perigee of the moon and perihelion of the Earth coincide; for the greatest tides in high latitudes, sun and moon should at the same time be at maximum declination. This can only happen near the solstices. This gives rise to a long interval of approximately 1670 to 1800 years[16] and between great maxima of the tides in northern waters. The last time there was a near approach to the exact alignment required to produce maximum tidal force in high latitudes was, according to the calculations used by Pettersson, in 1433 A.D. The Earth, sun, and moon were then in line, and the moon like the sun at maximum declination at the northern winter solstice.

4. Precession and Nutations of Earth Axis

The rotation axis of the Earth is subject to an oscillation forced by the regular variation of the pull of sun and moon upon the equatorial bulge, a pull which tends to tilt the polar axis of the Earth more, the greater the declination of sun or moon. The motion of the celestial pole of the Earth which results can be resolved into two components:

1. The progressive, nearly uniform motion of a fictitious average pole around the sun — called *precession*
2. The revolution of the actual pole of the Earth around the fictitious mean pole — called *nutation*

The nutation is nearly elliptical, though in detail many minor wobbles are superposed by the variations of lunar and solar distance due to the ellipticity of the orbits of moon and Earth. One nutation is completed in 18.6 years. The observed amplitude, which is only 9 sec of arc (about 300 m shift of the pole), is reduced by the inertia of the molten core of the Earth.

In addition, the polar axis of the Earth performs a free oscillation or nutation, in the course of which it describes from west to east a narrow angle, nearly circular cone. This free nutation with a period of about 14 months and an amplitude of about 1 sec of arc was first detected by Chandler in 1891[17] and is sometimes known as "Chandler's Wobble". The nutations of the pole of the Earth produce minute and unimportant variations of sea level, known as the *polar tide*. A 14-month periodicity has been demonstrated in them, though whether their phase in the different oceans can be related to the instantaneous position of the pole is not clear.

The cycles and changes described here do not allow us to explain any considerable long-term change of climate. We cannot exclude, however, a major astronomical disaster which hitherto has escaped our knowledge.

5. Effect of the Moon

The appearance of the moon by custom restricted to the particular phases of "new moon" when nothing is visible, "first quarter" when a semicircle is visible with the illuminated bow on the west, "full moon" when a full circle is visible, and "last quarter" when a semicircle is visible with the bow on the east. The changes of phase are caused by changes in the relative positions of Earth, moon, and sun. The moon rotates on its axis once in each orbital revolution and so the same face of the moon is always turned towards the Earth. The opposite side was only revealed to us in 1969 when the first astronauts landed on the moon.

The moon exerts a certain braking effect on the Earth which expresses itself by the phenomenon of the tides. It should however be noted that the tides are due to the combined gravitational forces of the sun and moon. This force prolongs daytime by 2 msec per century and removes the moon from the Earth by 4 cm/year. It is obvious that the 8-hourly friction of the tides with the ocean bottom brakes the rotation speed of the Earth. The "Alfred Wegener Symposium" which studied this phenomenon in 1980 in Berlin ascribes to the tidal force an effect of 4×10^{19} erg/sec, three quarters of it being due to the moon. However, by a retrograde computation of these data, one arrives at the paradoxical conclusion that the development of Earth in the Archeozoic era was thoroughly hampered by the extreme proximity of the moon to the Earth. During these early eras the moon tides would have upset the Earth crust by regular tidal waves of some dozens of meters. This could have easily produced far-reaching climate changes of the Earth.

B. Cosmic Interference

1. X-Ray Stars

Recently Hoffman[18] reported on cosmic X-ray bursts from various stars of the galaxy. They were discovered by rocket-borne X-ray detectors. The type of radiation emitted by a star depends mainly on the temperature of its surface. The surface temperature of our sun is about 5000°C. Slightly cooler stars appear red, hotter stars appear blue. At much higher temperatures, ultraviolet light and finally X-rays are emitted. Temperatures at which X-rays are emitted are millions to hundreds of millions of degrees, so X-ray stars must be vastly different from normal stars like our sun.

Riccardo Giacconi[19] of the American Science and Engineering Co. in Cambridge, Mass., conducted a rocket-borne experiment in mid-1962 and discovered a powerful celestial source of X-rays located in the direction of the constellation Scorpio. This X-ray source, called Scorpius X-1 (the first X-ray source discovered in Scorpio), delivers as many X-rays to the Earth as does our own sun, despite being tens of millions of times further away. Scorpius X-1 also emits normal light, but at such a low level compared to its prodigious X-ray output that it is visible only in fairly large telescopes. Clearly, this was a new kind of celestial object: an "X-ray star".

2. Pulsating Stars

In 1967, a balloon-borne X-ray telescope flown by a group led by Walter Lewin[20] of the Massachusetts Institute of Technology, observed Scorpius X-1 for about 2 hr. The data showed a fourfold increase in X-ray brightness occurring in less than 10 min. After this "flare", the brightness returned to normal in about 30 min. This was a hint of the strong variability which characterizes a great many of the more than 200 X-ray star sources known today. In addition to such flares, experimenters also discovered that sometimes a strong X-ray source could be detected during one rocket or balloon flight and then not be seen at all during a flight several months later. Such sources are called "transients" or "X-ray novae".

The American Science and Engineering group has now launched the Uhuru-X-Ray satellite from Kenya (Uhuru means "freedom" in the Swahili language). The first two such "X-ray binaries" to be discovered were Centaurus X-3 and Hercules X-1. These strange objects exhibited not only the variability in emission caused by eclipses — which occurred every few days — but also a much faster, regular variability every few seconds. This latter variability is believed to be caused by a beam of X-rays being swept around the sky like a celestial lighthouse by rapid rotation of the source. This type of behavior had first been observed in 1968 by radio astronomers from objects we now call "pulsars" (pulsating stars). Astrophysicists believe that pulsars are rapidly rotating neutron stars with strong magnetic fields. Neutron stars are incredibly dense objects left over after supernova explosions of stars.

3. Black-Hole Stars

The discovery of X-ray pulsars in binary systems has led to the development of a model believed to apply to many of the X-ray stars in our galaxy. In this model, a small dense object, probably a neutron star, but perhaps a black hole, orbits around a normal star. A black hole is formed in the collapse of a huge star and is so dense that not even light escapes from it. The gravitational field of the neutron star attracts matter from the atmosphere of the normal star, which is transferred to the neutron star and accretes onto its surface. As this matter falls onto the neutron star it gains kinetic energy, just as water does when it flows over a waterfall on the Earth. This energy turns to heat and raises the temperature of the falling material to millions of degrees — hot enough for the material to emit X-rays. To date, more than 30 X-ray burst sources have been discovered, both by the MIT SAS-3 group and also by the OSO-8 (8th Orbiting Solar Observatory) group at the Goddard Institute for Space Studies in Maryland. Their influence on our climate is not yet known. However, the possibility cannot be excluded that millions of years ago black-hole stars may have induced long-term climate changes on our Earth.

C. Solar Radiation
1. Present Period

For practical purposes, present solar radiation can be considered as constant. Yet, the possibility of great changes of solar radiation in geological time cannot be ruled out.[28] It has been shown that they agree with the observed climatic changes during the Quaternary. The first effect of an increase in the solar radiation would be to raise the temperature everywhere on the surface of the Earth, but more in low than in high latitudes. This would immediately increase the amount of evaporation from water surfaces and also the strength of the atmospheric circulation. More evaporation means more clouds and more precipitation. But clouds reflect back to space a large part of the solar radiation which falls on it. Hence an increase of cloudiness lowers the temperature. Another result of a large increase of solar radiation could be a slight rise of temperature and a great increase in cloudiness and precipitation. We now enter a long cool arid interglacial period, which lasts until a new increase of solar radiation brings about a renewal of the glacial and pluvial cycle.

At one stage during the cooling there may be a maximum snowfall and consequently glaciation. Further, each sun-flare outbreak diminishes the mass of the sun and consequently its normal radiation, leading to a general level of climate somewhat cooler than before. It seems impossible, however, that even a short-lived thousand-fold increase of solar radiation could have occurred without devastating animal and plant life, and there is certainly no trace of such a catastrophe in late Pliocene.

2. Sunspot Cycling

Sunspots show a very marked and persistent oscillation of approximately 11.2 years,

an annual mean of 10 or less at sunspot minimum and rising to a maximum
...s varied since 1749 from annual means of 46 to 154. This 11-year oscillation
...cognizable in the values of solar radiation, so it must represent a variation of
...her solar characteristic. Cycling seems to be an almost permanent characteristic
of the sun, for a cycle of 10 or 11 years has been found in the thickness of annual
layers in laminated deposits of various geological ages, including the Upper Carboniferous glacial clays of Australia.

There is little doubt that some relation exists between the sunspot cycle and terrestrial conditions, but it is very obscure. In tropical regions temperature averages about
1°C higher at spot minimum than at spot maximum; this relation disappears in temperate regions but reappears in the Arctic, where it is clearly shown both in the temperature of Spitsbergen and in the amount of ice in the Barents Sea, indicating that at
sunspot maximum the temperature falls in the Arctic and the area of the floating ice
cap increases. Over the world as a whole rainfall seems to have a tendency to be greatest at spot maximum, but there are many exceptions. One of the most interesting relations is between sunspot relative numbers and the frequency of thunderstorms, which
are most frequent at sunspot maxima.

3. Short-Term Climate Changes

Scientists have only recently come to learn how very sensitive the Earth is to small
changes in the flow of heat and other radiation from the sun, particularly if these
fluctuations persist. Recent numerical models of the atmosphere have shown that a
drop in solar flux of only 5% may be adequate to bring on a major glaciation. Even a
drop of 1 to 2% could initiate a "Little Ice Age" like one that brought extremes of
cold to Europe and North America in 1430 to 1850. In fact an enduring change in
sunlight of as little as one part in a thousand could produce a noticeable alteration in
climate, with social and economically important consequences.

That the sun was imperfect, spinning, and ever changing was to Renaissance philos-
ophers a shattering discovery. Immediate theological opposition to a blemished sun
was soon assuaged by the rationalization that sunspots were only clouds that floated
above a still perfect luminous surface. Closer examination, however, showed that they
were more intrinsic features. How serious a symptom sunspots really were was not
revealed until other, more active layers of the atmosphere of the sun were discovered
early in the 20th century. Only then was it realized that the visible surface of the sun,
the white photosphere where spots are seen with simple telescopes, is a deceptively
quiet layer, a thin veneer of stability that separates raging layers of the solar atmosphere. Hiding in the slowly changing sunspots are concentrated magnetic forces that
bring violence and disruption elsewhere on the sun. Beneath the placid, photospheric
surface, and largely hidden from view, lies a deep layer of boiling, churning turbulence. Sunspots and their patterns of coming and going result in part from the action
of this deep, convective layer, which can be far from constant. It is now suspected
that the 11-year sunspot cycle is merely a ripple in solar behavior compared with longer
and perhaps terrestrially more important changes on the sun.

Today the case is still unresolved, and some investigators still search for simple correlations between weather patterns and short-term changes on the sun. Many others
feel that a hundred years of controversial findings is enough to demonstrate that such
effects, if present at all, must be of minor importance or subtly connected to many
other, interacting influences. Modern knowledge of the complexity of the atmosphere
rules against the sort of direct, cause-and-effect relationships sought in the last century
and suggests that any mark of the sun on weather will be tangled in a twisted skein of
other atmospheric processes.

It is possible that solar connections have been sought in the wrong places or with the wrong index of solar activity. Most past efforts have looked for connections with visible solar activity, such as solar flares and sunspots. Yet these are only one manifestation of changes on the sun, and possibly not the most important. Observations of the sun from space have revealed significant fluctuations in the unseen ultraviolet and X-ray portions of the electromagnetic spectrum that emanates from the sun. It is also known that the sun bombards the Earth with a constant and irregular flow of charged atomic particles: electrons, protons, and the nuclei of heavier atoms. The speed and energy of this "solar wind" varies appreciably, but not in any simple relation to the sunspot cycle. Its properties could in some way trigger subtle weather change.

4. Long-Term Climate Changes

The case for important sun-weather connections becomes stronger when one moves from short-term weather to long-term climate. Recently Murray Mitchell of the National Oceanic and Atmospheric Administration (NOAA) and Charles Stockton of the University of Arizona Laboratory of Tree-Ring Research uncovered important new evidence from an extensive regional study of the annual growth rings of trees, whose widths vary with rainfall and thus offer a permanent record of annual moisture at each tree. They found that patterns of drought on the western plains of North America have recurred over the last 300 years in a clearly defined cycle of 20 to 22 years. These recurrent droughts, which include the "dustbowl" years of the early 1930s, the Plains drought of the early 1950s, and the Western drought of 1976 to 1977, mesh fairly well with the solar cycle throughout this time, falling at alternate minima of the 11-year sunspot cycle. Magnetic fields on the surface of the sun are known to reverse their magnetic polarity in the same period of 22 years, or two sunspot cycles, suggesting that the drought connection, if real, involves in some way the magnetic field of the sun, probably through the solar wind.[25]

Most striking in the historically accessible record of solar activity is a protracted minimum that lasted from about 1645 until 1715. Throughout this span of 70 years few sunspots were reported, auroras were infrequent, and the solar corona — the tenuous outer atmosphere of the sun seen during total eclipses — was less visible than it is today. It is not known whether the 11-year solar cycle was in operation during the time or not, but it is certain that the prolonged minimum in solar activity was real, for able astronomers with good telescopes regularly observed the sun. The remarkable absence of sunspots is well documented in books and journals of the day, and routinely mentioned afterward until the time of discovery of the sunspot cycle by Heinrich Schwabe, a German pharmacist. Thereafter, however, it seems to have been largely ignored or lost in the scramble to accept the notion of regular, cyclic behavior on the face of the sun.

Astronomers Gustav Spoerer in Germany and E. Walter Maunder of Great Britain's Greenwich Observatory revived interest in the anomalous period in papers published in the 1890s, but their efforts went largely unnoticed. Only recently, with new historical verification and additional evidence from tree rings, auroras, and other indirect sources, has it finally become accepted as a major feature of solar behavior. Modern scientists also know, as Spoerer and Maunder did not, that the 70-year period of sunspot absence coincided with a time of remarkable cold, a severe dip in the longer Little Ice Age, when temperatures of the Earth were colder than at any other time in the last 1000 years. This prolonged period of solar inactivity, now known as the *Maunder Minimum,* provided the first well-documented case of solar misbehavior and a yardstick for interpreting the longer, indirect record of solar activity that is found in the carbon-14 content of tree rings.

Tree rings hold the answers to whether and how carbon-14 production varied in the past. Carbon-14 produced in the upper atmosphere eventually finds its way as carbon dioxide into trees, where it enters the leaves through photosynthesis and becomes locked as cellulose in annual growth rings. From a knowledge of the age of a tree and from the chemical analysis of the wood in each of its dated rings scientists can recover carbon-14 variations and hence obtain a precisely dated history of the sun. It will be slightly ambiguous history, for factors other than the sun also regulate carbon-14 production, and it will fluctuate in time, for there are delays of 10 to 40 years between production of carbon-14 at the top of the atmosphere and its eventual assimilation into tree leaves at the bottom. The tree-ring record of carbon-14 leaves little doubt, however, about the solar signature of the Maunder Minimum; it is there — a dramatic increase in carbon-14 — in the right rings and for the right duration to verify the historical record of sunspot and auroral absences.

The Maunder Minimum, once it was clearly identified in both historical records and tree-ring carbon-14, was the "Rosetta Stone" that enabled scientists to read the much longer record of solar history found in the wood of long-lived trees like the bristlecone pine. In the long tree-ring carbon-14 record, which presently reaches more than 8000 years into the past, one can see the marks of repeated solar excursions like the Maunder Minimum. Each of these prolonged lows in solar activity corresponds to a similar period of anomalous cold on Earth, as derived from records of midlatitude glacier advance, as best as these are known. The correlation seems to confirm the more-than-chance connection between the Maunder Minimum and the cold excursion of the Little Ice Age.

Also evident in the tree-ring carbon-14 record are times of anomalously high solar activity, one of which is the present era, and each of these corresponds to times of warm climate and glacier retreat. The first drop in solar activity before the Maunder Minimum appears in both historical data from aurora and solar observations and in tree-ring carbon-14 between about 1400 and 1510 A.D. Called the Spoerer Minimum, it delineates the first of two severe dips of cold temperature during the Little Ice Age. Before that, evidence both in historical records and in tree-ring carbon-14 points to an anomalous high in solar activity that peaked in the 12th and 13th centuries A.D., a time known to climatologists as the Medieval Warm Epoch, when temperatures in Europe and America were last as warm as the present.

5. Future Trend

These remarkable coincidences between solar history and climate suggest that on time scales of 50 to 100 years changes in the sun may be a dominant force in altering climate. They indicate that for a long time scientists may have missed an important message in the pattern of sunspots and solar activity. Investigators were tuned for too long only to the sunspot cycle, which may be but the carrier wave. The amplitude modulation now seems to have carried the more important information on solar change and its effects on climate. In this slowly changing modulation there are hints of other cycles: one of about 80 years that has long been noted and a suggestion of a much longer one of about 2500 years, a period that also appears in records of glacier advance.

6. Solar Magnetism

The small part of the field of the Earth which is of external origin is produced by the entry into the high atmosphere of electromagnetic waves and particles emitted by the sun. Each of these produces characteristic effects on the magnetograms which thus provide a valuable continuous measure of solar activity.

The action of the solar-wave radiation is to ionize the high atmosphere and so make it electrically conducting. The sun and moon cause in the atmosphere tidal movements which, in the presence of the magnetic field of the Earth, induce, by dynamo action, electric currents in the conducting region. The magnetic field of these currents is observed at the ground, superposed on the main field of the Earth. The varying field produced by the thermal and tidal actions of the sun is, in general, clearly visible on the magnetograms as a characteristic local-time (S) variation of the magnetic elements. The purely tidal action of the moon on the atmosphere gives rise to a smaller varying field, the existence of which is demonstrated by arranging the magnetic element values according to lunar-time (L) variation. The amplitude of the solar daily variation of the elements, on other than highly "disturbed" days, is found to be a good relative measure of the ionizing wave radiation of the sun. The lower ionosphere, at 60 to 100 km, is the region mainly responsible for these magnetic effects.

The magnetic "disturbance" produced by solar particles is a highly complex phenomenon which is most frequent in the "auroral zones" at about 70° geomagnetic latitude where it is never entirely absent. At higher levels of disturbance, especially beyond the rather arbitrary lower level of a "magnetic storm", rapid field variations are world-wide and are accompanied by ionospheric disturbances, which produce anomalous radioreception, and by large Earth currents, which adversely affect cable telegraphy. At such times aurora is visible far equatorwards of its normal position. Field-strength fluctuations during large storms amount to about 3 to 8% of the undisturbed value, depending on latitude. In general, field direction changes are of a few degrees, but are much larger in high latitudes. They can surely explain the extreme climate changes during the ice ages.

Fundamental difficulties persist concerning the precise nature of the solar stream of charged particles which approaches the Earth, and the interaction of this stream with the main magnetic field of the Earth. Rapidly varying electric currents, concentrated mainly in the auroral zones, together with a "ring current" in the plane of the equator at a distance of several Earth radii, explain qualitatively many of the observed features of disturbance. There is good evidence that dynamo action in the high atmosphere is important in the production of climatic changes.

7. The Proton Effect

The penetration of solar protons into the magnetosphere at high latitudes has now been investigated directly with instruments on board the satellite HEOS 2, but the complexity of these processes is only likely to begin to be unraveled after the series of interrelated studies of the International Magnetosphere Study in the second half of the 1970s. Meanwhile, the best picture of short-term solar influences on the circulation of the atmosphere comes from the work of Roberts and colleagues,[24] which first linked the development of low pressure systems at high latitude with geomagnetic storms and ionospheric disturbances, and now includes observations of the influence of the solar magnetic sector structure on these phenomena.

The Wilcox[25] group concluded:

1. Meteorological responses tend to occur 2 or 3 days after geomagnetic activity.
2. Meteorological responses to solar activity tend to be most pronounced during the winter season.
3. Some meteorological responses over continents tend to be the opposite from the responses over oceans.

Since it was realized that the magnetic field of the Earth undergoes complete and, by geological standards, frequent reversals, several attempts have been made to link this phenomenon with the breaks in the fossil record which often occur at around the same time. Deep sea cores from several locations suggest the reality of this link, since during the past 2.5 million years eight species of radiolaria found in these cores became extinct, and six of these extinctions occurred in near coincidence with polarity reversals. One school of thought held that the link might operate through the mutagenic influence of the increased flux of cosmic rays penetrating to the surface of the Earth while the protecting magnetic field was absent during a reversal, but this now looks to be too small an effect to account for the changes; another speculation is that the changing magnetic field has a direct biological effect; but the most attractive theory is that climatic changes associated with the magnetic reversal play a part in producing the faunal extinctions.

8. Solar Ionization

Many solar flares produce large fluxes of particles, notably protons, which are detectable at the Earth; these events occur sporadically, although the general frequency of such solar proton events depends on the overall level of activity of the roughly 11-year solar cycle. The range of energies of individual particles produced in this way is from 10^4 eV to more than 10^9 eV, and the intensities and distribution of energy among the particles vary considerably from one event to another. Ionization produced by these particles in the stratosphere causes the formation of large quantities of nitric oxide, so that only a few intense events in a year will suffice to produce as much nitric oxide as is produced by direct oxidation of nitrous oxide. Through the interaction of nitric oxide with ozone this is directly affecting the efficiency of the ozone shield that protects the surface of the Earth from potentially harmful solar ultraviolet radiation. Although each solar proton event lasts only a few days, the lifetime of nitric oxide in the stratosphere is long, and its effect on the ozone layer is likely to persist for several years.

9. Nitric Oxide and Ozone Formation

The importance of ozone in the stratosphere for climate and the general circulation of the atmosphere needs no elaboration here, and even without developing this work further we already have a powerful new insight relevant to the occurrence of solar cycle effects in climatic variations. But it requires only a small further step to gain insight also into one aspect of the relationship between geomagnetism and climate.

Today, the presence of the geomagnetic field acts to steer the solar protons to high latitudes, so that the direct effects on the atmosphere are significant only above about 60° (either north or south of the equator). The nitric oxide is spread by atmospheric transport, but it is diluted by a factor of seven in proportion to the fraction of the surface of the Earth lying at latitudes above 60°. During polarity reversals, however, the geomagnetic field disappears, and is at best very weak for a few thousand years, during which time both solar protons and cosmic rays from outside the solar system can penetrate the atmosphere at low latitudes. In terms of the effect on ultraviolet transmission, the implications are clear. A large solar flare of the kind normally encountered during the solar cycle of activity would remove enough ozone to produce ultraviolet radiation levels 15% greater than those normal on the surface of the Earth today, while a flare 100 times larger than usual — a likely prospect sometime during a period of several thousand years — could lead to an increase in ultraviolet radiation at the surface of 160%. Such effects are likely to be significant hazards for some forms of life on Earth; even a flare 10 times larger than usual would increase the ultraviolet they receive by 55%.

Removing ozone and changing the transmissivity of the atmosphere are in themselves events of climatic importance, and it is possible that an increased flux of ions in the upper atmosphere could lead to increased cloud cover. The theories are, as yet, speculative; there is even room to suggest that the climatic effects produced as a result of the penetration of solar protons at all latitudes may have contributed to the faunal extinctions which occurred at the times of geomagnetic reversals, which certainly seem to have been exposed to additional stress factors. The idea applies equally to events more remote than those of the past 2.5 million years. For example, roughly one third of all living species became extinct at the close of the Cretaceous, which was a period marked by a resumption of polarity reversals, following a very lengthy period of normal polarity. The dinosaurs, it seems, may have been ill-adapted to cope with sudden bursts of ultraviolet radiation (or associated climatic events) to which they had been unable to adapt because the long absence of geomagnetic reversals failed to provide them with a taste of things to come.[26]

10. Solar Spots and Glaciation

Variations of rainfall in the temperate zone during the Christian era have run fairly parallel with the variations of solar activity shown by the records of sunspots and aurorae. But the phenomena of the Quaternary Ice Age were on a scale many times greater, and would require enormous and prolonged outbursts of sunspots, which seems quite improbable. Moreover, it has been shown that the terrestrial changes are not proportional to the sunspot relative number; the effect falls off rapidly as the relative number increases. Hence, while variations of sunspot activity may account for some of the minor glacial oscillations, it is unlikely that they played any appreciable part in the four main glacial advances and retreats of the Quaternary Ice Age, and still less likely that they caused that Ice Age as a whole. Therefore, the hypothesis assuming a relationship between sunspot increase and glaciation is not acceptable.

REFERENCES

1. Taylor, F. B., Bearing of the tertiary mountain belt on the origin of the Earth's plans (theory of 1908), *Bull. Geol. Soc. Am.*, 21, 179, 1910.
2. Wegener, A., *Die Entstehung der Kontinente und Ozeane,* Vieweg, Braunschweig, Germany, 1912.
3. Hess, H. H., History of ocean basins, in *Petrologic Studies,* Engel, A. E. J., James, H. L., and Leonard, B. F., Eds., Geological Society of America, New York, 1962.
4. Harland, W. B., Evidence of late pre-Cambrian glaciation and its significance, in *Problems in Paleoclimatology,* Nairn, A. E. M., Ed., Interscience, London, 1964, 119.
5. Gates, W. L., Report on Glacial Periods, U.S. National Research Council, Washington, D.C., 1974.
6. Gams, H., Aus der Geschichte der Alpenwaelder, *Z. Dtsch. Oesterr. Alpenvereins,* 68, 157, 1937.
7. Calder, N., *The Weather Machine,* Penguin Books, London, 1974.
8. Wood, E. S. , *Collins Field Guide to Archaeology,* 3rd ed., Collins, London, 1972.
9. Butzer, K. W., Some recent geological deposits of the Egyptian Nile Valley, *Geogr. J.,* 125, 75, 1959.
10. Delibrias, G., Ladurie, M., and Ladurie, L.E., Le fossil forêt de Grindelwald, nouvelles datations, *Ann. Économies, Sociétés Civilisation,* 1, 137, 1935.
11. Huntington, E., *The Pulse of Asia,* Houghton Mifflin, Boston, 1907.
12. Brooks, C. E. P., *Climate Through the Ages,* 2nd ed., Dover, New York, 1970.
13. Lamb, H. H., Britain's changing climate, *Geogr. J.,* 133, 445, 1967.
14. Marinatos, S., Life and art in prehistoric Thera, *Proc. Br. Acad.,* 57, 1, 1971.
15. Milankovitch, M., Mathematische Klimalehre, in *Hbch. d. Klimatologie,* Koeppen, W., and Geiger, R., Eds., Borntraeger, Berlin, Bd. 1, Teil A, 1930.
16. Pettersson O., The tidal force, *Geogr. Ann. (Stockholm),* 12, 261, 1930.
17. Chandler, C. S., On the variation of latitude, *Astron. J.,* 11, 83, 1891.
17a. Chandler, C. S., On the supposed secular variation of latitude, *Astron. J.,* 11, 109, 1891.
18. Hoffman, J. A., X-ray bursts, *N.Y. Acad. Sci.,* 17, 14, 1977.
19. Giacioni, R., Celestial X-rays, *N.Y. Acad. Sci.,* 17, 15, 1977.

20. **Lewin, W.**, Scorpius X-1 observations, *N.Y. Acad. Sci.,* 17, 16, 1971.
21. **Ladurie, L. E.**, *Times of Feast, Times of Famine,* Doubleday, Garden City, N.Y., 1971.
22. **Bryson, R. A.**, A perspective on climatic change, *Science,* 184, 753, 1974.
23. **Gribbin, J. and Lamb, H. H.**, Climatic change in historical times, in *Climatic Change,* Gribbin, J., Ed., Cambridge University Press, London, 1978, 68.
24. **Roberts, W. O. and Olson, R. H.**, Geomagnetic storms and wintertime 300 mb trough development in the N. Pacific and N. America area, *J. Atmos. Sci.,* 30, 135, 1973.
25. **Wilcox, J. M.**, Impact of solar activity, in Possible Relationships Between Solar Activity and Meteorological Phenomena, Bandeen, W. R. and Maran, S. P., Eds., NASA, Washington, preprint X-901-74-156.
26. **Gribbin, J.**, Astronomical influences, in *Climatic Change,* Gribbin, J., Ed., Cambridge University Press, London, 1978, 131.
27. **Flohn, H.**, Background of a geophysical model of the initiation of the next generation, *Quat. Res.,* 4, 385, 1974.
28. **Eddy, J. A.**, Climate and the changing sun, in *Encyclopedia Britannica Yearbook of Science and the Future,* Encyclopedia Britannica, Chicago, 1978, 145.
29. **Hollin, J. T.**, Interglacial climates and Antarctic ice surges, *Quat. Res.,* 2, 401, 1972.
30. **Frakes, L. A.**, *Climates throughout Geologic Time,* Elsevier, Amsterdam, 1979.
31. **Schwarzbach, M.**, *Alfred Wegener und der Drift der Kontinente,* Wissenschaftliche Verlagsgesellschaft, Stuttgart, 1980.

Chapter 5

MAN-MADE CLIMATE CHANGES

I. INTRODUCTION

A. History

Man's impact on the climate began 5000 to 9000 years ago as soon as he was able to alter the environment by burning and felling forests and tilling the Earth. It all began in river valleys — the cradles of ancient civilizations in arid regions, where Man learned to irrigate the soil and produce crops, i.e., vegetation, and where previously all but the river was dry, Man started to change the climate. The most extensive change wrought by Man prior to our own times was the gradual conversion of most of the temperate forest zone to crops, i.e., an artificial steppe or savanna.

Thus, until the industrial revolution, and probably until the present century, Man has had little effect on the climate except on a very local scale. Up to the middle of the present century, outside the limits of urbanized and industrialized areas and away from artificial lakes and irrigated areas, Man's effects on the climate have hardly been significant in comparison with the magnitude of the natural fluctuations of climate.

Anxieties are now, however, reasonably felt, and widely expressed, about the possible imminence of global effects owing to the recently introduced energy at Man's disposal, the increasing variety of pollutants, his ingenuity in developing new pollutants, and his rapidly increasing technical power. Certain grandiose schemes for modifying world climate deliberately, e.g., inducing a permanent warming by abolishing the arctic ice, with the aim of increasing the total cultivable area of the globe, have been discussed from time to time and assessments made of the energy cost and technological development required to achieve them. Aside from these aspects, a basic objection to all such plans is that no changed pattern of climate (which either Man could devise or Nature might produce) would be an improvement for all human needs. Indeed, some would be bound to produce distress, and probably disasters, in some regions. None could properly, or advisably, be pursued without the broadest international agreement, and all would be liable to entail some movement of populations. Moreover, it cannot be pretended that the consequences of deliberate action would all be predictable.

Schemes have also been published for deliberate actions to change the global pattern of climate as a strategy of war. It has been reported (*The Times,* London, June 24, 1975) that talks have begun on a Soviet proposal for a draft treaty, submitted to the United Nations, under which the nations would forswear attempts to manipulate the weather or climate as an instrument of war.

B. Methods of Human Intervention

Within the present century various other local, and mostly short-term, measures for weather protection have been introduced. These range from special localized heat sources for fog clearance on airfields, and smoke sources for frost protection of orchards and vineyards, to "cloud seeding" — with silver iodide or "dry ice" (solid carbon dioxide at very low temperature). This would provide condensation nuclei for ice crystals and encourage the transfer of supercooled water droplets in the clouds to the ice nuclei — thus inducing rain or averting a risk of hail damage by preventing the growth of large hailstones within the cloud. A useful review of these techniques, their economics and effectiveness is given by Maunder[1] with references to the extensive literature on such practices. None of these measures, however, can be thought to have any large-scale or lasting effect on the climate.

Proposals for large-scale modification of the climate of wide regions or of the entire globe have so far not proceeded beyond the research and discussion stage, though several meetings of experts have been held. Such proposals were among the items considered by the Conference on Modification of Climate, organized by the Voejkov Geophysical Observatory and the Institutes of Applied Physics and Geography of the U.S.S.R. Academy of Sciences at Moscow, April 25 to 28, 1961[2] and by the Panel on Weather and Climate Modification of the U.S. National Academy of Sciences in 1966.[1,10]

C. Large-Scale Schemes for Climate Modification

To many people the prospect of Man's meddling with the large-scale climate of the Earth seems impious. To the extent that this attitude is based on apprehension of the likelihood of conflicting interests and of the limits of predictability of the actual effects of climate, it is justified indeed. Certainly, there is an obligation to proceed humbly and with caution, since any climatic change or fluctuation must introduce a new variability of agricultural yields in some areas, and world population has now reached a level at which the food reserves built up over the past 30 years have been largely exhausted, e.g., partial harvest failures in the Soviet grainlands in 1972 and in India in 1972 and 1974. It is hard to see how any contrived shift of the large-scale climatic pattern — even one aimed at increasing the total cultivable area of the world — could be achieved without some local or regional disasters (Malthusianism).

1. Creation of Inland Seas in Africa

Bergeron[3] proposed at the Conference on Physics of Precipitation held at the Woods Hole Oceanographic Institution, in Massachusetts, in 1959, to increase the rainfall of the semi-arid (Sahel-Sudan) zone of North Africa, the savanna and the steppe, by "injecting" extra humidity on a large-scale into the monsoon air. Thus, Bergeron's scheme might be brought into operation in line with a plan originally conceived by the German engineer Soergel[4] in the 1930s. Suitably massive damming of the Congo River at the Stanley Gorge near Kinshasa (Leopoldville), at a point where in the Cretaceous era the river broke through the rocky rim of an earlier lake, should in the course of a few years cause the great inland basin of Zaire to fill with water to a level sufficient to turn back the tributary river Ubanga. The river would then flow north to join the Shari River and fill Lake Chad once more into an inland sea about 2 million km^2 in extent. The two water bodies together would cover an area of about 3 million km^2 or 10% of the African continent.

The meteorologically important aspect of the plan is that this great expanse of water would lie broadly within the regions that are for much of the year covered by the cloud masses and vertical air circulation of the equatorial rain system, so that the extra water available should be continually recycled through the vegetation growth in the zone and the atmospheric circulation over it. By contrast, the water lost by evaporation from the new Nasser Lake at Aswan under the usually cloudless skies at that latitude is mainly transported far beyond the confines of Africa before it falls to the Earth again.

2. Creation of a Siberian Sea

According to Flohn,[4,10] the proposal of the Russian engineer M. Davidov to dam the rivers Ob and Yenesei (Jenesei) should produce an expanse of water of about 250,000 km^2 in northwest Siberia. From this sea, it is suggested, water could be channeled through the Turgai region to arrest the drying up of the Caspian and Aral Seas and water the arid steppes in the region, while still allowing enough flow over the dams to maintain navigation on the lower reaches of the Ob and Yenesei.

3. Removal of Ice on the Arctic Ocean

Much research has been devoted in the Soviet Union to the possibility of removing the Arctic sea ice and the probable effects of such action on the climates of the northern landmasses, as well as to the question of whether the ice would form again.[5-7a] This may be considered the biggest project proposed for deliberate modification of climate. The object would be to raise the temperature of northern lands and so to open up great regions to human settlement and cultivation.

An alternative plan, proposed by the Russian engineer Borisov[8] and much publicized, is to dam the Bering Strait, blocking off the present flow of water from the Pacific Ocean into the Arctic and instead pumping 500 km² water per day from the Arctic to the Pacific. Borisov's supposition is that thereby more, and warmer, Atlantic water would be drawn into the central Arctic than now reaches there.

Of all the large schemes mentioned in this section Bergeron's[3] is the proposal that seems soundest, in that it takes most strategic advantage of the natural tendencies of the energy pattern and organization of the global wind circulation. For this very reason it can be criticized as the least radical plan, in that it seeks only to extend and enlarge somewhat the already existing rainfall distribution. Indeed, it may do so on a sufficiently large-scale (in a warm climate zone where the moisture-carrying capacity of the air is very great) to make a very significant contribution to food production. Moreover, Bergeron has suggested that the project should be preceded by smaller scale tests in partly similar air-mass situations in southeast Australia, southwest Norway, or elsewhere. All the other proposals listed are open to serious objections because of the likelihood of undesirable and very wide-ranging side effects, some of which might well nullify the main objectives of the schemes.

II. INADVERTENT MODIFICATION OF CLIMATE

Since about 1950 many studies have been undertaken, and a great variety of evidence has been presented, showing that modern man-made modifications of the environment may affect climate not only locally but also on a regional, or even global, scale. The observed variations of average temperature over the world since 1880 have sometimes been interpreted as man-made effects, despite the fact that similar — and also greater — variations occurred before the industrial era, attributable only to natural influences such as we have discussed in this book. Several recent reports[9,10] have concluded that man-made effects on regional or world climate may now be barely detectable, but must be expected to wax in importance, even allowing for the buffering and damping influence of the oceans with their great storage capacity for carbon dioxide and water.

A. Gases
1. Carbon Dioxide
a. Facts

The increase of CO_2 in the atmosphere from the burning of fossil fuels is estimated to be now 13% above the amount present in 1850. When considered in conjunction with estimates of the amount of fossil carbon burned, it appears that only 50 to 75% of the carbon dioxide produced has remained in the atmosphere, the rest having entered the oceans and the biosphere. For many years it has been recognized that variations in the atmospheric carbon dioxide concentration could result in climate changes.[19]

Interest in the CO_2-climate problem has increased in recent years, possibly for the following reasons:

1.   Recent measurements show a continuing increase in CO_2 concentration.

2. Improved climate models substantially confirm the earlier predictions of the climatic effects of increasing concentration of atmospheric CO_2.
3. Deforestation may be a significant source of atmospheric CO_2, comparable with fossil fuel combustion; if this is so, considerable modification is required to existing models of atmosphere-ocean CO_2 exchange.
4. High carbon isotope results from tree ring studies, although inconclusive, suggest changes in atmospheric composition which are not explained by fossil fuel CO_2 input alone.
5. The environmental impact resulting from the continued use of fossil fuel as an energy source should be of no less concern than that resulting from alternative energy sources.

b. Trends

Projections of the carbon dioxide in the atmosphere, based on extrapolation of Man's birth rate and fuel consumption, suggest by conservative estimates a rise to 22% above the 1850 level by 1990 and to 32% by 2000 A. D.

c. Climatic Response

It is generally recognized that changes of a few degrees in global mean temperature would have a very significant effect on society in general and agriculture in particular. It should be mentioned that a 2°C warming would return the Earth to a condition similar to that which existed during the climatic optimum 6000 years ago. In the long term, such changes may actually be beneficial to some areas of the globe, but in the short term it is this rate of change that would impose great climatic stress on society.

It seems unlikely that the biosphere as a whole has or will in the near future, respond significantly to increases in atmospheric CO_2 levels, despite the well-known effects of CO_2 fertilization on photosynthesis in laboratory and glass-house conditions. Often, under field conditions, other environmental factors, such as the availability of trace elements or water, limit growth response. The annual variation of CO_2 concentration measured at Hawaii is believed to represent the seasonal uptake and release of carbon by the Northern Hemisphere terrestrial biota. As the amplitude of this variation has not changed significantly over 15 years there is no evidence of a significant change in the annual turnover of carbon. This means that vegetation now existing in deforested areas has about the same annual exchange as the forest it replaces, or that higher atmospheric CO_2 concentrations have increased annual exchange sufficiently to balance the loss of forested areas. Probably the most significant effect upon the biosphere of increased CO_2 will arise from the indirect effect of long-term rainfall and temperature changes.

d. Stable Carbon Isotopes

A significant input to the atmosphere of nonfossil fuel CO_2 has occurred due to deforestation. The underlying principle of this and related research is that the carbon in each of the major reservoirs (namely the ocean, atmosphere, and the biosphere-fossil fuels) may be identified by its isotopic composition. If a significant input of carbon occurs from one of the reservoirs to the atmosphere, then there will be a change in the isotopic composition of the atmosphere.

It may be that stable carbon isotope variations in the annual growth rings of trees are not directly related to past atmospheric composition. For example, under circumstances where atmospheric mixing is not sufficiently intense, photosynthesis, which itself is a fractionating process, will locally deplete the air of ^{12}C. Further photosynthesis would therefore result in the uptake of carbon isotopes in a ratio unrelated to that of the free atmosphere. There are also uncertainties about the temperature dependence of the photosynthetic fraction, while the equilibrium distribution of isotopes between the atmosphere and oceans may also depend on temperature. Finally, the marked difference in isotope composition between cellulose and lignin may introduce variations in growth ring isotope ratios, where these components of the wood are not treated separately.

The study of stable carbon isotopes, particularly in tree rings, is in its infancy. Thus, as was the case in the area of dendrochronology, we need to obtain sufficient experience to assess which trees (from specific areas) can be used as indicators of past atmospheric composition.

e. Oceanic Uptake of CO_2

Many attempts have been made to model the carbon cycle and in particular to describe the role of the oceans in the uptake of fossil fuel CO_2. It may be tentatively concluded that the deep sea seems the most likely sink for the excess CO_2 produced by man.

In addition, a physical understanding of feedback mechanisms existing between climate changes and large-scale atmosphere-ocean CO_2 exchange is required in any realistic model. For example, with overall atmospheric warming (implying a greater temperature increase at the poles than at the equator), meridional sea surface temperature gradients decrease, and the vertical thermal stratification of the oceans increases. Such changes might then tend to reduce the efficiency of transport of CO_2 from the atmosphere to the oceans (as well as reduce the available nutrients in surface water) and increase the proportion of injected CO_2 remaining in the atmosphere.

Carbon dioxide is taken up in the oceans by reaction with carbonate (CO_3^{--}) to produce bicarbonate (HCO_3^-) ions, i.e., CO_2 (diss.) $+ CO_3^- + H_2O = 2HCO_3$. The capacity of ocean water to take up CO_2 therefore depends on a source of carbonate ions. Such a source is the dissolution of calcium carbonate ($CaCO_3$). However, in general, ocean water is supersaturated in CO_3^{--} with respect to $CaCO_3$ and it is not expected that dissolution of even a fraction of the vast ocean sediments of $CaCO_3$ will occur until atmospheric CO_2 concentrations have increased by about fourfold. It has been suggested, that at this point the survival of marine organisms possessing carbonate skeletons may be threatened, although it is not certain that the ability of such organisms to precipitate carbonate necessarily depends on a high degree of carbonate supersaturation.

Ocean surface water CO_3^{--} does, however, appear to be in equilibrium with magnesium calcite and it has been suggested that this may represent an additional source of carbonate ions and thus influence the capacity of the oceans to take up CO_2. This also requires further study.

f. Climatic Forecast

Probably coal gasification and liquefaction will be used to avoid major fuel shortages in the next few decades, with CO_2 continuing to be released into the atmosphere in ever-increasing amounts. Is it possible, therefore, to suggest ways in which the accumulation of CO_2 in the atmosphere can be avoided? This question requires considerably more attention than it has received thus far. The oceans, if freely mixed, would be capable of taking up some 80 to 90% of the excess CO_2. Brief consideration was given to the possibility of trapping CO_2 out of power plant exhaust systems and piping it directly into the deep ocean. Alternatively, the oceans could be fertilized with nutrients that at present limit productivity, and thus increase the sedimentation of carbon into the deep ocean. In both cases, the schemes appear to be prohibitively costly, requiring additional use of fuel, and may be associated with further undesirable environmental effects.

Realistically, one might argue (at least) for a reduction in the rate of consumption of fossil fuels, with probable reduced environmental effects in the long term. This could be achieved by a reduction in the effective amount of energy consumed per person, with the implied change in attitudes by society as a whole, and the encouragement of reduced population growth. We must avoid any further trend towards the depletion of the global biomass, including deforestation, the drainage of wetlands, and the exposure of high organic matter soils to erosion.

In summary, there is no immediate practical solution to the future problems likely to arise from increasing atmospheric CO_2 concentrations and the probable climatic consequences predicted by the models. This imposes a responsibility on the scientific community to actively pursue the relevant research, and present evidence in the context of future energy policies. The direct evidence as might occur from observations of climatic change itself will not be clear for some time because of the variable nature of climate. We need to explore practical means of reducing the accumulation of CO_2 in the atmosphere, and give serious consideration to suitable plans of action relating to the climatic effect on agriculture, economics, politics, and society in general.

On balance, the effect of increased carbon dioxide on climate is almost certainly in the direction of warming but is probably much smaller than the estimates which have commonly been accepted. Thus, there does not seem to exist any threat to our climate by carbon dioxide increase.

2. Infrared Absorbing Gases

Marked increases in surface temperature can be produced by large-scale releases of carbon dioxide and the chlorofluorocarbons, the effect being due to their ability to absorb infrared radiation in the atmospheric "window", and their long persistence. This suggests that we should be alert to the buildup of any other trace gases that have similar properties, and there are a great many of them.

One such trace gas is nitrous oxide, which is mainly maintained at its present concentration of about 0.28 ppm by biological decay and conversion processes taking place in soil and in the oceans — processes referred to as "denitrification". It has been suggested that the increasing use of nitrate fertilizers by mankind may accelerate the biological production of nitrous oxide and raise its atmospheric concentration, with implications for both surface temperature increase and stratospheric ozone concentration decrease. The amount of this increase in nitrous oxide concentration is still uncertain, since estimates vary from a trivial increase to as much as a factor of 2 in the early part of the next century. The latter would poduce a warming in the order of $0.5°C$, but this may be considered as an estimate on the high side until we understand the global nitrogen cycle better, and specifically the relative productions by ocean and soil biota.

B. Particulate Matter in the Atmosphere

1. Trend of Increase

The atmosphere's total burden of dust, smoke, and other particles (about 1970) has been estimated by Mitchell[11] at about 4×10^7 tons, of which about 1×10^7 tons may have come directly or indirectly from Man's activities. The history of these particles is much less well established than that of CO_2, but the man-made and man-induced contribution is probably increasing, perhaps by as much as 4% yearly, which would lead to a concentration of 60% above the 1970 level by 2000 A.D.

Man-made particles are estimated to be ten times as numerous in the atmosphere over the Northern Hemisphere as over the Southern Hemisphere. Note the large contribution of particles from agricultural practices as compared with industry: much of this is attributed to the smoke particles from the burning of great areas of grass (savanna fires) each year in the dry season. There must also be an increased uptake of mineral dust in dry areas where the natural vegetation has been destroyed.

According to Mitchell[11a] about 90% of the particle load on average is within the troposphere. The remaining average figure of 10% in the stratosphere is primarily volcanic dust and sulfate particles in the Junge Aerosol layer at about 20 to 25 km, also largely of volcanic origin; this item varies by up to 100 times from one group of years to another, depending on volcanic activity.

2. Effect on Climate

The effect on climate of the dust in the lower atmosphere is hard to assess, and there is as yet no firm consensus or conclusion about it. The effect on radiation passing through the atmosphere depends on the particle sizes and varies according to the wavelength. Particle sizes between 0.1 and 5 μm are certainly the most important because of their considerable abundance and relatively long residence, provided they are not washed out by rain. Average residences of 10 days in the lowest 1.5-km zone up to 30 days in the upper troposphere are suggested, compared with 1 or 2 years for the volcanic dust particles in the stratosphere. Such particles are much more effective in intercepting solar shortwave radiation than the outgoing long-wave radiation from the Earth. Thus the Earth's albedo is increased. The net effect in any area must depend on whether the dust is lighter or darker (has higher or lower albedo) than the ground beneath.

Pending more adequate observational evidence, it seems safe to conclude that the effect of man-made dust in the lower atmosphere is mainly in the direction of cooling, at least in all the lower latitudes, and there is probably a tendency to moderate the diurnal and annual extremes of temperatures.

3. The Cooling Effect

Periods of global cooling have been recorded over the past two centuries after major volcanic eruptions spewed tons of dust particles into the air. Along with world population, the amount of dust in the atmosphere has doubled since the 1930s, despite the absence of major volcanic eruptions. Some scientists fear that increased amounts of atmospheric dust may act as insulation, reflecting the sun's rays away from the Earth and lowering temperatures. While it is too soon to assess the global climatic significance of Landsberg's[12] estimates, studies of the glaciers indicate that high concentrations of dust in the air were associated with the onset of the last ice age.

4. The Warming Effect

While rising atmospheric dust levels may be cooling the Earth, another pollutant, carbon dioxide, appears to be exerting a warming influence through the so-called "greenhouse effect", which slows the release of the heat of the Earth into space. The effect is analogous to that which has been supposed to operate in a greenhouse, whereby the surface of the Earth is maintained at a much higher temperature than the temperature appropriate to balance conditions with the solar radiation reaching the surface of the Earth. The atmospheric gases are almost transparent to incoming solar radiation, but water vapor and carbon dioxide in the atmosphere strongly absorb terrestrial radiation emitted from the surface of the Earth and reemit downward radiation.

Another effect of CO_2 especially in combination with SO_2 pollution of the air is its dissolving effect on marble structures. The wonderful statues like the Acropolis in Athens and those of Venice bear witness of this devastating change of urban climate.

The warming effect is supplemented, in urban and industrial areas, by thermal pollution or waste heat. While small compared to the heat the Earth receives from the sun, waste heat around large cities can markedly alter local climates. Urban areas average several degrees warmer and have less snowfall and more rain than adjacent rural areas. In Washington, D.C., the frost-free growing season for backyard gardeners is now 1 month longer than in outlying areas. The urban heat island created by the city of Paris causes temperatures to average higher at the center of town than at Le Bourget airport on the outskirts of the city, meaning more fog and rain for Parisians.

5. The Drying Effect

As population pressures increase, desire for more rainfall to improve agricultural production intensifies; accordingly, the weather is intentionally modified to suit human needs. The seeding of rain clouds, a means of bringing rainfall to dry areas, has become an expedient response to short-term drought situations. Unfortunately, cloud seeding often does no more than cause rain to fall in one place at the expense of another, and thus presents still another potential source of international tension.

6. The Blurring Effect

Airborne dust is probably the most common and easily recognized of the man-made pollutants that affect climate. City rains that leave dark blotches on clothes are evidence of extremely dirty, dust-filled air.

Humans generate airborne dust in every daily activity from suburban driving to tilling the soil. Around large cities with millions of individual actions each generates its own small quantity of dust, the particles often act as cloud seeders, increasing rainfall. In a study of the St. Louis metropolitan region, a group of researchers from the University of Illinois found rainfall as much as 30% higher in Edwardsville, Ill., downwind of St. Louis and the Alton-Wood River industrial area, than in surrounding areas.[9]

7. Aerosols

Aerosol particles can both scatter and absorb sunlight, and they also absorb and reemit infrared radiation to a limited extent. When a particle scatters solar radiation some of the scattered radiation will be directed upward as well as downward, and the upward component will be lost to space. This results in less sunlight reaching the ground and an increase in the net albedo, or reflectivity, of the atmosphere-Earth system, which would cause a net cooling. However, when a particle absorbs some of the solar radiation it heats the particle and the air around it, and the effect of this is to reduce the net albedo. Theory tells us that in order to decide whether aerosols cause an increase in the net albedo cooling or a decrease in warming we must take into account the ratio of the particle absorption to its backscatter. When aerosols are over a dark surface, such as the ocean, they are more likely to increase the net albedo than when they are over a light surface, such as a snow field or a low cloud deck—or over land generally.

There has been a widely shared belief that such aerosols generally cause a cooling, the argument being that when spread evenly around the Earth their effect over the dark oceans is to increase the albedo and thereby prevent some of the sunlight from being absorbed by the Earth-atmosphere system. Recently, however, it has been pointed out that most of these antropogenic aerosols exist over the land, where they are formed, and that they are sufficiently absorbing to reduce the albedo rather than increase it.

Aerosol production in each country is proportional to gross national product, and the aerosols drift with the surface winds and remain in the atmosphere with a mean residence time of 5 days before they are rained out or washed out or directly deposited at the surface. Their distribution is very uneven, being mostly confined to the industrialized regions of the Northern Hemisphere, though a certain portion does drift out over the Atlantic and Pacific Oceans, and a considerable part of Europe's "gross national pollution" drifts over North Africa, particularly in the wintertime. These aerosols absorb more solar radiation than natural aerosols, and their extinction values are generally large enough to cause a lowering of the albedo of the atmosphere-Earth system over the land, and thereby cause a warming. However, largely due to our lack of quantitative knowledge of the optical characteristics of these aerosols and their distribution, we cannot yet assign any number to this warming effect.

There are other effects that aerosols may have on the climate of a region, especially its rainfall, their role as condensation and freezing nuclei, which may be significant. Another effect that may be significant is their influence on the stability of the lower layers of the atmosphere. Since they absorb a certain amount of solar radiation, the upper part of a low-lying aerosol layer will be warmed, and the absorption and scattering processes will both cause a decrease in the solar radiation reaching the ground. The result is a warming of the upper part of the aerosol layer in daytime and a decrease in the rate of warming at the ground, and this causes the stability of the atmosphere near the ground to be larger than it would be in the absence of the aerosol particles.

Aerosols, probably because they are so very obvious to the eyes of the population in large cities, have been the subject of vigorous attempts to control them. The result is that in many cities of the world the aerosol content, particularly of larger particles, has shown a definite decrease. The same cannot in general be said for the total aerosol content of the atmosphere observed in Europe and the eastern U.S., where secondary aerosol production of smaller submicron particles, especially sulfates from the sulfur dioxide produced by burning high sulfur fuels, have become a dominant factor in regional air pollution. It should be noted that the practical problems posed in these regions by the ecological and health effects of increasing quantities of sulfate particles probably far outweigh their influence on the regional climate.[18]

C. Water Vapor Discharged by Aircraft

There is no doubt that the frequency of condensation trails near major routes has increased the amount of cirrus cloud there. But the percentage of the Earth surface so affected is still small, and evidence of any significant effect on global cloudiness is still inconclusive. It must be important to make use of satellite technology to establish (for the first time) how much the total cloudiness of the Earth varies from year to year from both natural and man-made causes. It is supposed on theoretical grounds[13] that an increase of water vapor in the stratosphere will tend to warm the climate at the Earth's surface.

D. Pollution of the Stratosphere

The still minute water vapor content of the stratosphere seems to have increased significantly over recent years, but this is thought to have been a natural variation. If supersonic aircraft and rocket exhausts add H_2O to the stratosphere in the future, perhaps up to 3 to 3.5 ppm by 1990, there may be a significant effect on world temperature. The effect will be a balance between that due to increased albedo of the Earth (if the frequency of thin clouds at the mesopause level of 80 km increases) and the absorption and reradiation of the long-wave radiation of the Earth. It is generally supposed that the net result would be a tendency towards warming at the surface of the Earth.[14]

E. Ozone Danger

The amount of ozone in the stratosphere also increased during the 1960s. This too is supposedly a natural fluctuation, though its causes and mechanism are as yet unknown (see Volume I, Chapter 3).

Other forms of pollution of the stratosphere, due to Man, have been discussed by the W.M.O. Executive Committee Panel on the Meteorological Aspects of Air Pollution which in 1970[9] expressed concern lest chemical constituents of jet aircraft exhausts cause the destruction of the stratospheric ozone layer. Chlorofluorocarbons, popularly known as freons, used as a propellant in aerosol sprays are also believed capable of destroying the ozone in the stratosphere. There is certainly no evidence of that happening, at least in the years surveyed until now, and no specific process has been alleged whereby the ozone would be likely to be destroyed. But since ozone is highly active chemically as an oxidizing agent, the fear of what might happen if the volume of stratospheric flying increased greatly and introduced significantly more foreign constituents into the atmosphere at those levels is not unreasonable. Caution and careful observation, preferably continuous monitoring of the stratospheric ozone, are clearly advisable.

F. Desert Formation

Charney[15] has given theoretical justification for the view that overgrazing in the Sahel region of the southern fringe of the Sahara may have increased the natural tendency to desertification. Owing to the high albedo of desert sands, estimated by Charney from satellite measurements as averaging 35%, destruction of the vegetation means an increased loss of heat from the surface, which encourages sinking motion (anticyclonic subsidence) in the atmosphere over the regions of bare surface — a "positive feedback" mechanism which tends to intensify and perpetuate the dryness.

G. River Diversion

Very serious are the possible side effects that have to be considered in connection with schemes to divert the flow of the great rivers which enter the polar basin to water the arid lands in middle latitudes in North America and Asia. These are illustrated by the Soviet scheme for central Asia. There is no doubt that the water emergency which threatens central Asia is real. Current use of the two central Asian rivers Amu Darya and Syr Daria to irrigate 5,500,000 ha results in half the water being lost by evaporation, etc., before it reaches the Aral Sea, the level of which is reported to have dropped consequently 1.5 m in recent years.

There are 34,000,000 ha of fertile land in the region which, if irrigated, could yield cotton, rice, sugar beet, fruit, and pasture and which probably must be brought into cultivation within the next 50 years to meet the needs of an increasing population. Current and projected needs would, it is estimated, cause the total disappearance of the Aral Sea unless further sources of water can be arranged.

The level of the Caspian Sea is also dropping: it is reportedly by now a continuous fall, at least partly owing to the use of the water from the rivers which supply it — mainly the Volga — for irrigation. Total inflow from the rivers into the Caspian Sea over the 20 years from 1946 to 1965 is estimated as 900 km² less than it would have been but for its use for irrigation. The Caspian is receiving 30 km² too little water yearly to balance the evaporation from its surface, and this deficit is expected to increase by 50 to 70% by 1985. The decreasing volume of water in the Caspian Sea, which has no outlet and is shallow, particularly in its northern part, increases its salinity and threatens the valuable fishing industry there, the main source of supply of the sturgeon which provides the famous caviar.

The salt content of the Sea of Azov, described as the most productive fishing basin in the world, is also likely to increase as water from the Black Sea flows in to make up the declining flow from the rivers Don and Kuban due to industry and irrigation.

H. Summary of Heating Effects

There are several anthropogenic causes for climatic change in terms of their effects on the mean temperature of the surface. To a first approximation they can probably be considered as additive, because they represent small fractional changes (and therefore follow more or less linear relationships, though for larger changes some of the effects must definitely become nonlinear).

The conclusion to be drawn from this is that the anthropogenic influences on global mean surface temperature, especially the effects of carbon dioxide and fluorocarbons (unless the use of the latter is severely restricted), will probably begin to dominate over natural processes of climatic change before the turn of the century and will result in a decided warming trend that will accelerate in the decades after that. It is quite likely, in fact, that this warming trend has already begun.[18]

III. MAN'S DIRECT OUTPUT OF HEAT

Artificially generated heat is significant so far only on a local scale, in the urban heat islands, where the surface air temperature in the city center may average up to 2°C higher than in the surrounding countryside and in certain industrialized reaches of rivers, lakes, and inlets of the sea. Annual average temperatures in such waters (near power stations, steelworks, oil refineries, etc.) are occasionally 5 to 10°C above those prevailing in neighboring unaffected water.

Budyko[16] and Galtsov[2] have suggested many times that Man's output of heat is likely to become the most serious cause of change of climate within about 100 years, and mentioned that countermeasures such as the deliberate creation and maintenance of an artificial dust veil to reduce the receipt of radiation from the sun may have to be undertaken. Schneider and Dennett[17] have pointed out in the same connection the advantage of tapping natural energy flows such as wind and water movement, which would not add heat to the environment.

In summary, the various ways in which Man has polluted the atmosphere and disturbed the environment so far have probably not individually affected the general level of prevailing world temperature by more than 0.1 to 0.2° C. As some pollutants cause warming and others cooling of the Earth, their overall effect on world temperature has probably been very small, though anxiety about the continuing carbon dioxide increase remains.

The effect of human activity — both positive and negative — on world temperature, tends towards exponential increase. Forward estimates of its net effect are therefore unreliable, but it appears likely that the increasing output of man-made heat will gain the upper hand in the next century unless stronger controls are instituted than so far contemplated, lest there be a catastrophic breakdown of the international economy.

IV. CLIMATE MONITOR

The unstableness of our climates has induced Professor Hubert H. Lamb of the University of East Anglia, Norwich, England, to establish a Climate Research Unit. The "Climate Monitor" is the official organ of this unit. It was established in 1972 with the principal aim of improving knowledge of the long record of changes of world climates, and among others, examining possible scientific approaches to the problem of formulating advice on future climate periods. Special attention is focused on agricultural, industrial, and commercial planning by governmental authorities. The present climatic variations point to the possibility of future global cooling and necessitate our careful watch of the many factors involved in its origin and prevention.

A review of climatic history during the last 18,000 years (since the last glacial maximum) demonstrates the variability of our climate. This can be fully understood only when taking into account a multitude of nonlinear "internal" interactions between atmosphere, ocean, ice and snow, soil, and vegetation. Furthermore one has to consider some "external" causes of climatic variations, e.g., variations of the solar radiation, volcanic eruptions, which are at present unpredictable. Then man-made effects — slow changes of surface albedo, air pollution, increasing rise of CO_2 content and of other trace gases adding to a "greenhouse effect" — should not be neglected.

Holding the natural effects constant, increasing global warming will depend on the sum of anthropogenic effects. A scenario of climatic evolution under these assumptions has been presented in this book.

V. CONCLUSIONS

In the present monograph we have attempted to present the many factors which are involved in short-term and long-term changes of climate. They can affect our health and environment by their connection with the manifold forces by which nature governs weather and climate. The reader will hopefully have become aware of the fact that in short-term changes of climate the electrical changes in the air — ions, sferics, and electrofields — have decisive effects on one third of the world population, a fact which has not so far been duly appreciated. Another third suffers from heat or from hot dry winds, a fact well known to anyone who lives in an area harassed by heat or winds of ill repute. There still remains the final third of mankind, who do not mind any of these adverse effects of nature, and to these people this presentation may not yet appeal because they are still young and resistant. However, they may have learned how many of the people around them suffer from meteorological phenomena. They may also have discovered that such sufferings can be alleviated today by modern treatments, which comprise drugs as well as ions and electrofields.

On the other hand long-term changes of climate can only be guessed by the mosaic of changes showing up in detective-like methods of dating, such as radiometry, amino acids, thermoluminescence, dendrology, pollen rain, lake varves, volcanic layers, ice sheets, ocean sedimentation, geomagnetism, archeology, and astronomy. As we are dealing here with aspects of millions of years, interpretation of changes varies with the different authors. Still there emerges the fact of seven ice ages which were brought about by pole wandering and continental drift. Man has survived all these hardships and we can only hope that our interference with the present climate will not change our future one. Various factors can produce man-made climate changes, such as carbon dioxide enrichment, increase of particulate matter in the atmosphere, vapors discharged by industry, aircraft and traffic, pollution of the stratosphere, ozone enrichment, desert formation, river diversion and dams, commercial drying of lakes, and construction of artificial fish ponds.

Whereas long-term changes of climate are mainly concerned with cold periods, short-term changes had to be devoted mostly to disturbances caused by heat and drought which poses a menace to our food problems. On the other hand, there are ample means today to protect ourselves against cold, as the example of Alaska, Canada, and Scandinavia has shown, but we have not yet discovered means of combating the heat which harasses us outdoors.

Against all the evil forces of nature many tactics have been tried. Yet, successfully countering the vicious power of cold and ice still needs special new devices, as the customary ones broke down when Napoleon tried a winter in Russia, and Hitler followed in his footsteps. The problem of forecasting hurricanes has been successfully tackled, but their prevention seems to be still very remote. The same holds for heatwaves and their fatal impact.

This book has been written in a country where the deluge and the impact of hot dry desert winds have even found their place in the Bible, but where, on the other hand, the generally favorable climate has allotted to this part of the world the privilege of being the cradle of three religions. It seems that the development of religious thought, of literature and of civilization, needs a temperate or subtropical climate, and perhaps the intuition furnished by the forces of nature, weather, and climate.

Charles Dudley Warner (1829 to 1900) once said: "Everybody talks about the weather but nobody does anything about it." We trust that our work in this field has now shown that something can be done about it, and that modern treatment can relieve everyone of weather ailments and sufferings.

REFERENCES

1. **Maunder, W. J.,** *The Value of the Weather,* Methuen, London, 1970, 388.
2. **Galtsov, A. P.,** Conference on the modification of climates, *Izvestia Seria Geogr.,* 5, 128, 1961.
3. **Bergeron, T.,** Possible man-made great-scale modifications of precipitation climate in Sudan, *Geophys. Monogr.,* 5, 399, 1960.
4. **Soergel, H. (1930) in Bergeron, T.,** Opening address at the Int. Conf. on Cloud Physics Tokyo and Sapporo 24.5-1.6.1965, *Meddelande,* 91 (Met. Inst. Kungl Univ.) Uppsala, 1965.
5. **Budyko, M. I.,** The heat balance of the Earth, in *Climatic Change,* Gribbin, J., Ed., Cambridge University Press, London, 1978, 85.
6. **Doronin, J. P.,** On the problem of eradicating the arctic ice, *Probl. Arktiki Antarktiki,* 28, 21, 1968.
7. **Rakipova, L. R.,** Changing the climate by artifical action on the ice of the Arctic basin, *Glav. Uprav. Gidromet. Sluzb, Met. i Gidr.,* 9, 28, 1962.
7a. **Rakipova, L. R.,** Variation in the zonal distribution of temperature as a result of climate modification, in *Sovremennye Problemy Klimatologii,* Budyko, M.I., Ed., Leningrad Gidrometeoizdat, 1966, 358.
8. **Borisov, P. M.,** The problem of the fundamental amelioration of climate, *Izvestia Vses. Geogr. Obshchestva,* 94, 304, 1962.
9. World Meteorological Organization, Climatic Fluctuation and the Problems of Foresight, Unpubl. Final Rep. Working Group Comm. Atmos. Sci., Geneva, 1972.
10. National Academy of Sciences, Understanding Climatic Changes: A Program for Action, Panel on Climatic Variation, U.S. Committee for the Global Atmospheric Res. Progr., National Research Council, Washington, D.C., 1975.
11. **Mitchell, J. M.,** A reassessment of atmospheric pollution as a cause of long-term changes of global temperature, in *Global Effects of Environmental Pollution,* Singer, S. F., Ed., Reidel, Dordrecht, Holland, 1973.
11a. **Mitchell, J. M.,** The effect of atmospheric aerosols on climate, with special reference to temperature near the earth surface, *J. Appl. Meteorol.,* 10, 703, 1971.
12. **Landsberg, H. E.,** Planning for the climatic realities of arid regions, in *Urban Planning for Arid Zones: American Experiences and Directions,* Golany, G., Ed., John Wiley & Sons, New York, 1978, chap. 2.
13. **Machta, L.,** Stratospheric water vapor, in *Man's Impact on the Climate,* Mathews, W. H., Kellogg, W. W., and Robinson, C. D., Eds., MIT Press, Cambridge, Mass., 1971, 410.
14. **Kukla, G. J.,** Recent changes in snow and ice, in *Climatic Change,* Gribbin, J., Ed., Cambridge University Press, London, 1978, 113.
15. **Charney, J. G.,** Dynamics of deserts and drought in the Sahel, *J. R. Meteorol. Soc.,* 101, 193, 1975.
16. **Budyko, M. I.,** Some ways of influencing the climate, *Met. i. Gidr. MGA 14A-43,* 2, 3, 1962.
17. **Schneider, S. H. and Dennett, R. D.,** Climate barriers to long-term energy growth, *Ambio,* 4, 65, 1975.
18. **Kellogg, W. W.,** Global influences of mankind on the climate, in *Climatic Change,* Gribben, J., Ed., Cambridge University Press, London, 1978, 203.
19. **Pearman, G. I.,** The carbon dioxide-climate problem, *Clean Air,* 11, 21, 1977.

GLOSSARIES

I. METEOROLOGICAL TERMS

Adiabatic changes — change of temperature due to pressure without heat entering or leaving the system (*a* = not, *diabaino* = pass through).

Aerosol — a colloidal system in which the dispersed phase is composed of either solid or liquid particles, and in which the dispersion medium is some gas, air, or fog (*aer* = air, *sol* = from solution).

Albedo — reflecting power of a surface (*albedo* = whiteness).

Altocumulus — high white or gray cloud (*altum* = height, *cumulus* = heap). *A. Castellanus* — if resembling a castle.

Altostratus — atmospheric pressure distribution in which there is a high central pressure relative to the surroundings. It proceeds towards a low-pressure center, the cyclone (*cycle* = circle).

Anvil Cirrus — cloud resembling an anvil at its top.

Beaufort Wind Scale — a system of defining 12 speeds and strengths of winds: (calm) light air, light breeze, gentle breeze, moderate breeze, fresh breeze, strong breeze, near gale, gale, strong gale, storm, violent storm, hurricane.

Bishop's Ring — a dull reddish-brown ring around the sun due to diffraction of fine volcanic dust in the high atmosphere.

Black body — a planetary radiation of 250 K (Kelvin units) which absorbs all wavelengths and emits also at all wavelengths — except light.

Buys Ballot's Law — when a "High" advances towards a "Low" the pressure exchange does not proceed in a straight line due to Earth rotation eastward. It follows that if you stand with your back to the wind the lowest pressure lies on your left in the Northern Hemisphere and on your right in the Southern Hemisphere.

Cirrocumulus — cloud in form of heaped hair locks.

Cirrostratus — cloud in form of a hair-lock layer.

Cirrus — cloud in form of a hair-like band or patch (*cirrus* = hair lock).

Convection — a mode of heat transfer within the atmosphere involving the movement of hot air upwards and cold air downwards.

Coriolis force — the acceleration which air possesses by virtue of the Earth's eastward rotation, with respect to axes fixed in the Earth.

Cumulonimbus — cloud in form of a rainy heap.

Cumulus — detached clouds simulating rising mounds, domes or towers, the upper part resembling a cauliflower (*cumulus* = heap).

Cyclone — atmospheric pressure distribution in which there is a low central pressure relative to the surroundings. In the middle and high latitudes a cyclone is only a depression or Low, but in the tropics it develops into a storm of great intensity (*cycle* = circle).

Depression — low atmospheric pressure (Low, cyclone).

Diabatic process — permeating wind or radiation.

Doldrums — equatorial oceanic regions of light variable (mainly westerly) winds, accompanied by heavy rains, thunderstorms and squalls. These belts, variable in position and extent, have a systematic north and south movement some 5° either side of a mean position, following the sun with a lag of 1 or 2 months.

Equatorial low-pressure belt — the quasi-continuous belt of low pressure lying between the equator and the subtropical high-pressure belts of the Northern and Southern Hemispheres. Here humidity is so high that slight variations in stability cause major variations in weather.

Front — the transition zone between two air masses of different density and temperature, typical for weather changes.

GMT — Greenwich mean time.

Greenhouse effect — the heating effect created by the atmosphere upon the Earth by virtue of the fact that the humid atmosphere absorbs and reemits infrared heat radiation.

High — an area of high atmospheric pressure. It causes anticyclonic wind circulation.

Humidity, Relative (R.H.) — the percentage ratio of the actual vapor pressure of the air to the saturation vapor pressure. Naturally, it is high at dawn and sunset and reaches a minimum in the afternoon. This value is of practical interest to meteorologists and physicians.

Inversion — a departure from the usual decrease or increase of temperature with altitude. A typical occurrence is the formation of two warm layers above the surface with a cold layer in between.

Ionosphere — portion of the Earth's atmosphere, extending upwards from about 60 km to an indefinite height, which is characterized by a concentration of ions and free electrons high enough to cause reflection of radio waves.

Ions — electrically charged atoms or molecules which create an electrically conducting medium. They move quickly in the atmosphere (*ion* = going).

Isobar — lines of the synoptic chart which connect points of equal atmospheric pressue (*iso* = equal, *baros* = weight).

ITCZ (Intertropical Convergence Zone) — a relatively narrow equatorial zone in which air masses and winds originating in the Northern and the Southern Hemisphere converge.

Katabatic wind — descending mountain breeze.

Low — an area of low atmospheric pressure, also called a depression. It causes a cyclone movement.

Millibar (mbar) — a pressure unit of 1000 dyn/cm² introduced for reporting atmospheric pressures: 1 bar = 10^3 mbar = 10^6 dyn/cm², roughly corresponding to 750 mm mercury (*mille* = a thousand, *baros* = weight).

Monsoon — the term originally referred to the winds of the Arabian Sea which blow for about 6 months from the northeast and for the other 6 months from the southwest. It is now used specially for the Indian southwest winds which bring rain (*mausim* = season).

Nimbostratus — cloud in the form of a rainy layer.

Occlusion — lifting of an air mass by the coalescence of a cold front meeting a warm front. It results in the formation of a cold front on the surface which shuts off the warmer air and keeps it above.

Orography — topography of an elevated terrain. Orographic lifting means therefore the passage of an air current over a mountain chain (*oros* = mountain, *graphe* = write).

Rotor — a turbulent cloudy wind between two mountain barriers around an axis parallel to the range.

Savanna — the term applied to a type of tropical climate, with a wet and a dry season, in which the most common form of vegetation is the tall tropical grass called "savanna grass".

Smog — smoke-fog, a fog in which smoke or atmospheric pollutants combine to create an aerosol, i.e., a dispersion of small particles in foggy air.

Stratocumulus — cloud in the form of a flattened heap.

Stratosphere — that region of the atmosphere, lying above the troposphere and below the mesosphere, in which, in contrast to these regions, temperature does not decrease with increasing height. It extends from the tropopause to a height of about 50 km (*stratos* = layer).

Subsidence — the slow downward motion of the air over a large area, usually connected with adiabatic warming up of the subsiding air.

Sunspot cycle — a cycle with an average length of 11.1 years, but varying between about 7 and 17 years.

Synoptic chart — a map representing the distribution of selected meteorological elements over a large area at a specified instant of time. It mainly shows barometric pressure, temperature, clouds, precipitation, and dew.

Trade winds — trade winds blow in a steady direction thus furthering shipping and commerce. The name derives from "tread" (Fr. *trade,* It. *strada,* Gm. *Passat)* and was only later changed into "trade". They are the tropical easterlies which diverge from the subtropical high-pressure belts, centered at 30 to 40° latitude N and S, towards the equator.

Tropopause — the atmospheric boundary between the troposphere and the stratosphere (*tropos* = turning point).

Troposphere — the lower layer of the atmosphere extending to 9 to 16 km above the Earth. In this region the decisive changes of temperature, precipitation, and clouds take place. The upper limit reaches the tropopause (*tropos* = turning point).

Trough — a low-pressure feature of the synoptic chart indicating the descent from a High to a Low, formerly called a V- shaped depression.

Weather front — see Front.

II. MEDICAL TERMS

Adrenal glands — a hormone-producing organ situated above the kidney. It produces adrenaline and noradrenaline in its core (medulla) and the corticosteroids in its shell (cortex). The latter are excreted in the urine as 17-KS and 17-OH (*ad* = beside, *ren* = kidney).

Adrenaline = Epinephrine — a hormone secreted by the adrenal medulla in response to splanchnic stimulation, and stored in the chromaffin granules, being released predominantly in response to hypoglycemia. It is also produced synthetically. It is the most powerful vasopressor substance known, increasing blood pressure, stimulating the heart muscle, accelerating the heart rate, and increasing cardiac output (*ad* = beside, *ren* = kidney).

Creatinine — a basic substance produced by protein metabolism. It indicates daily protein turnover and appears nearly always in fixed quantities in the urine (*creas* = meat).

Diuresis — the daily amount of urine (*dia* = passing urine).

Endocrine glands — they work by secretion that is not discharged by a duct from the body, but is given off into the blood and lymph, taking an important part in metabolism, and regulation of sex functions (*endo* = internal, *crine* = secreting).

Epinephrine — cf. adrenaline (*epi* = above, *nephros* = kidney).

Epiphysis — cf. pineal gland (*epi* = above, *physis* = growth).

5-HIAA (5-Hydroxyindole Acetic Acid) — a chemical constituent of the urine created by metabolic destruction of serotonin, indicating its production and turnover (*indole* = indigo derivative).

Histamine — a hormone producing allergic reactions, occurring in all animal tissues. It is a powerful dilator of the capillaries and a stimulator of gastric secretion (*histos* = tissue, *amine* = ammonia derivative).

Homeostasis — a tendency to uniformity or stability in the normal body states (internal environment or fluid matrix) of the organism. Its rules were established by W. B. Cannon (Boston) (*homeo* = uniform, *stasis* = standing).

Hormone — a chemical substance, produced in the body, which has a specific effect on the activity of a certain organ; originally applied to substances secreted by various endocrine glands and transported in the bloodstream to the target organ on which their effect was produced. The term was later applied to various substances not produced by special glands, but having similar action (*hormo* = stimulate).

17-Hydroxysteroids (17-OH) — the compounds formed by cortisone and similar adrenal cortex hormones which are secreted in the urine and indicate cortisone production and turnover (*hydroxy* = chemical configuration containing hydrogen: H and O; oxygen).

Hypophysis — cf. pituitary (*hypo* = below, *physis* = growth).

Hypothalamus — the central master gland of the brain. It secretes hormones called "releasing factors" which stimulate secondary hormone production in the pituitary and contains our most important centers which regulate sex, growth, thyroid, adrenal, visceral activity, water balance, temperature, pain, sleep, and emotion (*hypo* = below, *thalamus* = brain center).

17-Ketosteroids (17-KS) — steroids which possess ketone groups on functional carbon atoms. The 17-ketosteroids have a ketone group on the 17th carbon atom. They are found in the urine of normal men and women and in excess in certain adrenal cortex and ovarian tumors (*keton* = chemical configuration containing an *O* = oxygen, *steato* = fat, *oid* = shaped).

Neurohormones — hormones stimulating neural mechanisms (*neuron* = nerve, *hormo* = stimulate).

Noradrenaline = Norepinephrine — a hormone secreted by the adrenal medulla in response to splanchnic stimulation, and stored in the chromaffin granules, being released predominantly in response to hypotension (*nor* = a chemical formulation of nitrogen: N, *o:* ohne = without, *r:* radical = chemical substituent).

Norepinephrine — cf. noradrenaline.

Ovary — the female sex gland in which are formed the ova and the female sex hormones, estrone and progesterone (*ovum* = egg).

Pancreas — a huge intestinal gland which produces an enzyme-rich juice to digest food, and also two hormones: insulin which lowers blood sugar and glucagon which elevates it (*pan* = all, *creas* = flesh).

Parathyroid glands — four tiny endocrine glands embedded in the thyroid gland which regulate the calcium metabolism (*para* = beside, *thyreos* = shield, *eidos* = shape).

Pineal gland — a cerebral body which is sensitive to light and heat. It regulates by hormonal secretion timing of our hormonal functions (*pineal* = pine-cone shaped).

Pituitary — the endocrine headquarters. It is situated underneath the hypothalamus, protruding from the brain and produces special hormones, the tropins, which regulate the activity of our endocrine glands (*pituita* = mucus).

Reticular formation — a cerebral area of net-like structure connecting the brain cortex with the hypothalamus and the spinal cord.

Serotonin — a hormone serving as neuro-transmitter in the brain. It is also produced in the enterochromaffin cells of the intestines and transported from there into the whole body where — when released — it may produce manifold irritant reactions (*serum* = blood fluid, *tonus* = pressure).

Stress — the sum of all nonspecific biological phenomena elicited by adverse external influences, including damage and defense. It may be localized, as in the local adaptation syndrome (LAS), or systemic, as in the general adaptation syndrome (GAS). The importance of stress in life and disease was recognized by Hans Selye (Montreal).

Testes — the male sex glands in which are formed the semen, and the hormones — testosterone and androsterone (*testis* = witness who stands by).

Thymus — a lymphoid gland situated under the breast bone which regulates immunity. It is highly developed in the newborn and degenerates with progressing age (*thyme* = plant in the shape of the gland).

Thyroid — the gland which covers our throat and neck like a shield. It produces hormones which regulate metabolism, heart action and mental acuity (*thyreos* = shield, *eidos* = shape).

Thyroxine — an iodine-containing hormone of the thyroid gland. It promotes metabolism, stimulates heart and pulse, and promotes mental activity (*thyr* from thyroid, *oxine* = oxidated hormone).

III. GEOLOGICAL TERMS

Algonkian epoch — the youngest stratigraphic division of the Precambrian epoch, named after the Canadian Indian tribe.

Alluvium series — a stratigraphic term for detrital material transported by a river and deposited — usually temporarily — at points along the flood plain of a river. It is commonly composed of sands and gravel. Many important ore minerals, e.g., gold, platinum, diamonds, are locally found concentrated in alluvial deposits.

Archeozoic period — primary life period, designating the first geological period dating 1000 to 3000 million years ago.

Azoic epoc (Greek = no life) — a name given to Precambrian strata presumed to have been formed before life appeared on Earth. The name is of doubtful validity and is rarely used.

Boreal (Latin = northern) — implying a northern element in a fauna. Since the term originated in the Northern Hemisphere it is effectively synonymous with "cold", and is often used with reference to an arctic element in a fauna, especially in Tertiary and Quaternary stratigraphy.

B.P. — before present, pertains to dating the present period from 1952 when radio-isotope dating was introduced by Libby.

Cambrian epoch — (from the Roman name for Wales — Cambria, as proposed by Adam Sedgwick, a British geologist, in 1835), is the oldest system of rocks in which fossils can be used for dating and correlation. The period commenced at least 530 ± 40 million years ago, and had a duration of at least 70 million years. Generally speaking, the base of the Cambrian shows a marked unconformity with the underlying sediments, and contains the first unequivocal shelled fossil remains, although in some places there is evidence that Cambrian sediments accumulated in a basin of sedimentation formed in Precambrian times, with little or no disconformity between the two. While most groups of invertebrates are represented in the Cambrian, only a few are sufficiently abundant to be of any geological importance. Trilobites (Arthropoda insects) were abundant, especially the more primitive forms. Brachiopoda (lampshade-like mollusks) were common and inarticulate forms predominated throughout. The vertebrate Graptolites Chordata first appear in the Upper Cambrian.

Carboniferous epoch — a system named (by Conybeare, 1822) from the widespread occurrence of carbon in the form of coal. It extends from 345 to 280 million years ago, lasting 65 million years. The Carboniferous is perhaps the most important system economically, containing as it does the bulk of the world's coal reserves, together with important deposits of iron ore, oil shale, oil, fire clay, and ganister, a siliceous rock used to line furnaces.

Cenozoic period — recent life period, designating the geological period extending from the beginning of the Tertiary to the present.

Chalcolithicum — copper age, from *chalco* (Greek) = copper (not chalk!). Nine parts of copper with one part of tin produce a rather hard and durable alloy called bronze which characterizes the tools of this epoch 2000 to 3000 years ago.

Continental drift — a conception generally linked with the names of F. B. Taylor and A. Wegener. According to this hypothesis, continents now separated were once joined together as, for example, South America with Africa, and North America with Europe; in general, changes in both latitude and longitude are implied.

Cretaceous epoch — (Greek *creta* = chalk). The Cretaceous period has a duration of approximately 72 million years, from 136 to 64 million years. The name was adopted by the Belgian J. J. D'Omalius D'Halloy in 1822 and introduced into England by Fitton. The division of the period into Lower and Upper Cretaceous is at about 100 million years, each of these main divisions being further subdivided into six stages. The Cretaceous is represented on all continents and is a very important period from the viewpoint of volume of sediments and amount of exposures. The fauna of the Cretaceous is closely related to that of the Jurassic. Ammonites (Mollusca) were abundant, especially in the lower part, but became rarer in the later layers, and finally became extinct at the end of the period. They are used as zone fossils wherever possible. Echinoderms (Echinodermata) and lamellibranchs (Mollusca) were locally important and are used as zonal indexes in the absence of ammonites. Belemnites (Mollusca) also became extinct during the period. Brachiopoda (mollusk-like shells) flourished but suffered a marked reduction in numbers at the end of the period. On the whole, corals (Coelenterata) were distinctly less abundant than in previous eras. On land, the dinosaurs continued to be dominant, but became extinct at the end of the period; Cretaceous mammals are known but are insignificant in size and numbers. Advanced flowering plants (Angiosperms) became important in the Cretaceous.

Devonian epoch — a system named after the county of Devon in southwest England as proposed by the British geologists Sedgwick and Murchison in 1839. It extends from 395 to 345 million years ago, having a duration of 50 million years. The marine Devonian is zoned on early ammonoid cephalopods (Mollusca), whereas the Old Red Sandstone is zoned largely on brackish or freshwater fish. No new fossil groups are represented in the marine Devonian, but some of the groups common in the lower Paleozoic sediments became rare or extinct. In the Old Red Sandstone fish, plants, and freshwater mollusks are the only common forms of fossil found.

The fish underwent considerable development, starting with primitive, armored, jawless forms which developed into advanced jawed forms, and which, at the top of the Old Red Sandstone, developed lungs and were on the threshold of an amphibious mode of life. Plants developed initially as unspecialized swamp forms rarely exceeding 60 cm in height, which later evolved to give rise to the large tree-ferns which appeared at the close of the period.

Diluvium — a stratigraphic term formerly used for what are now referred to as superficial deposits, supposedly the desposits from the Noachian deluge. The term was subsequently used for any flood deposit and later for any superficial deposit.

Eocene — early recent epoch at approximately 40 to 60 million years ago (Greek: *eos* = dawn, *kainos* = recent). Producing only 1 to 5% of modern species.

Evaporite — a sediment resulting from the evaporation of saline water. A typical sequence is as follows: (1) potash and magnesium salts, (2) rock salt (halite), (3) gypsum or anhydrate, (4) calcite and dolomite. The importance of the evaporites for human industry and agriculture is obvious.

Holocene period — entire recent period from approximately 100,000 years ago to present. (Greek: *holos* = whole, *kainos* = recent). The denotation was coined by the British geologist Sir Charles Lyell (1797—1875).

Ice Ages — glacials made up of alternate glacial and interglacial stages or periods. The best known ice ages are (1) the Huronian in Canada, occurring very early in the Proterozoic; (2) the Precambrian and early Cambrian which occurred in the early Paleozoic (about 530 million years ago) and left traces widely scattered over the world; (3) the Permo-Carboniferous, occurring during the late Paleozoic (from 275 to 225 million years ago) and which was extensively developed in northern India and the northeastern U.S.; and (4) the Quaternary or Pleistocene which began about 1 million years ago and may not yet have ended.

Jurassic epoch — named after the Jura Mountains of France and Switzerland (by Von Humboldt in 1795), the Jurassic is the period of time extending from 195 to 135 million years ago, lasting for 60 million years. It is divided into the Lower Jurassic (or Lias), the Middle Jurassic (or Dogger), and the Upper Jurassic (or Malm); the boundaries are dated at 172 and 162 million years, respectively. The Jurassic has been further subdivided, on a very minute scale, into a number of local subdivisions. Over 100 fossil zones have been erected, grouped into 11 stages, the lower limit being the preplanorbis zone, while the upper limit is marked by the zone of Cypridea punctata, Arthropod. The fauna of the Jurassic was extremely diverse and varied, the chief members being the ammonites (Mollusca), by which the period is zoned, the hexacorals (Coelenterata), and the sea urchins echinoids (Echinodermata). Brachiopoda (Lamp shells) were abundant. Lamellibranchs (Oysterikes) and gastropod snails (Mollusca) were also abundant, many forms being associated with coral reefs. In addition to the ammonites, the cephalopods (squids) were also represented by the belemnites (Mollusca). The dominant terrestrial animals were the dinosaurs, which reached their maximum size and distribution in the Jurassic and occupied most of the ecological niches. The first birds appeared in the Upper Jurassic, but the mammals, although present, were an insignificant element of the fauna, being rarely larger than the modern rat. The flora included many forms still existing at the present day, such as cycads with pinnate leaves, ginkoes with fan-shaped leaves, conifers (evergreen trees), and ferns (spore bearers).

Mesolithicum — Middle Stone Age designating an epoch in which man knew only the use of primitive flint tools 10,000 years ago.

Mesozoic (lit. "middle life") — an era ranging in time from 230 to 70 million years ago — a duration of 160 million years. It comprises the Triassic, Jurassic, and Cretaceous epochs. The Mesozoic was preceded by the Paleozoic and followed by the Tertiary. In older works the Mesozoic is sometimes referred to as the "Secondary era". The most spectacular elements of the Mesozoic fauna were the giant reptiles, the first mammals appearing at the commencement of the era. The chief invertebrates are Mollusca, Brachiopoda, and Echinodermata.

Miocene — less recent epoch at approximately 0 to 25 million years ago (Greek: *meion* = less, *kainos* = recent) denoting the time when 29 to 60% of modern species were produced.

Neolithicum — New Stone Age designating an epoch in which polished stone implements were used. The name was coined by John Lubbock (later Lord Avebury, 1865) and is used for a time 3500 to 5000 years ago.

Oligocene epoch — little recent epoch at approximately 25 to 40 million years ago (Greek: *oligos* = little, *kainos* = recent), denoting the time between the Eocene and Miocene epochs. It produced 10 to 15% of modern species.

Ordovician epoch — named after the Ordovices — an ancient Celtic tribe of Central Wales — as proposed by Charles Lapworth, an English geologist, in 1879. The period extended from 500 to 435 million years ago, a duration of 65 million years. Widespread vulcanicity characterized this epoch and the onset of the Caledonian orogeny is evident. Compared with the Cambrian, more advanced trilobites (Arthropoda) became abundant and the Brachiopoda are represented by articulate forms. Crinoids (Echinodermata) became abundant and the first tabulate and rugose corals (Coelenterata) appeared. The first vertebrates (fish) appeared in North America, but have not yet been recorded in Europe. The most important fossils are the graptolites (Chordata) by means of which the system is zoned.

Orogenesis — mountain formation (Greek: *oro* = mountain, *genesis* = creation).

Paleocene epoch — ancient epoch at approximately 600 to 63 million years ago (Greek: *palaios* = ancient, *kainos* = recent). During this time animal species were not yet developed.

Paleozoic period — term used for the period ranging from 600 to 230 million years ago, a duration of 370 million years. It comprises the Cambrian, Ordovician, and Silurian in the older or Lower Paleozoic sub-era, and the Devonian, Carboniferous, and Permian epochs in the newer or Upper Paleozoic sub-era. The boundary between the Lower and Upper Paleozoic is drawn at 400 million years. The Paleozoic was preceded by the precambrian and followed by the Mesozoic. In older works, it may sometimes be referred to as the "Primary era". Its lower boundary is marked by the appearance of the first fossil trilobites (Arthropoda). The Lower Paleozoic fauna is characterized mainly by invertebrates, including trilobites, graptolites (Chordata), and brachiopods (Brachiopoda). The earliest fish remains occur in the Ordovician. In the Upper Paleozoic, graptolites and trilobites became extinct, while corals (Coelenterata) and crinoids (Echinodermata) increased in abundance. Fish were important in the Devonian; amphibia and reptiles developed in the Carboniferous. The first terrestrial flora appeared in the Devonian.

Permian epoch — named at the suggestion of Murchison in 1841 after the province of Perm in Russia. It is the period of time from 280 to 225 million years ago, a duration of 55 million years. It marks the end of the Paleozoic. Because of the widespread occurrence of continental conditions during late Carboniferous, Permian, and Triassic times, defining the lower and upper limits is often difficult. Where marine deposits occur, the incoming of Pseudoschwagerina (a large Foraminifera— Protozoa) marks the base. The period marked the extinction of a number of fossil groups, the most important being the trilobites (Arthropoda) and the tabulate and rugose corals (Coelenterata). The only new group to become widely established were the reptiles, which were the first vertebrates to sever their association with water. Floras also showed a marked change in the Permian, the large, primitive forms of the Carboniferous trees being largely replaced by the more advanced conifers. Fossils are extremely rare, the most common being tracks and trails of the early reptiles.

Pleistocene period — (most recent epoch) at approximately 100 to 1000 thousand years ago. (Greek: *pleistos* = most, *kainos* = recent), a misnomer because the Pleistocene preceded the Holocene by about 100,000 years. It produced most of our present species.

Pliocene epoch — more recent epoch (Greek: *pleion* = more, *kainos* = recent). Again a misnomer because it happened 1 to 10 million years ago, and produced 50 to 90% of modern species.

Pole wandering — continental drift divided a central continent called Pangaea into America and Europe plus Africa. The rift — still visible at the bottom of the Atlantic — induced pole-wanderings and tilting of the Earth axis. The North Pole shifted from its present position to Malaya and back via Siberia and Europe to the Arctic. The South Pole shifted via South America to South Africa and thence in a great arc across Australia back to Antarctica. The change of temperature must have been instantaneous since the mammoth carcasses regularly found in Siberia contain food rests in their stomach and their flesh is palatable.

Precambrian epoch — the system of time from the consolidation of the Earth's crust to the base of the Cambrian. The terms Proterozoic, Azoic, and Archean have been used either as synonyms or partial synonyms. Because of the unconformable relationships and metamorphic state of much of the Precambrian, correlation on more than a local basis has proved exceptionally difficult. The advent of radioactive dating has done much to clear up numerous problems and the nucleus of a world-wide correlation scheme is beginning to appear. The duration of Precambrian time is probably not less than 4000 million years ago, and during this time a number of orogenies are known to have occurred. Most Precambrian rocks, therefore, can be shown to have undergone one or more Precambrian orogenies, as well as post-Cambrian ones. However, relatively unmetamorphosed sediments of Precambrian age are known from a number of areas. In some of these (e.g., in South Australia and Leicestershire) Precambrian fossils have been described, but their affinities are obscure.

Primary era — an obsolete name for a time which had its name changed from the Precambrian, subsequently the Precambrian plus Paleozoic, later to the Paleozoic only.

Proterozoic period — anterior life period pertaining to the geological period preceded by the Archeozoic.

Quaternary era — the term for the latest system of time in the stratigraphic column, 0 to 2 million years ago, represented by local accumulations of glacial (Pleistocene) and postglacial (Holocene) deposits which continue, without change of fauna, from the top of the Pliocene (Tertiary). The Quaternary appears to be an artificial division of time to separate prehuman from posthuman periods. The Quaternary is marred by the appearance of the ice age. It can be safely assumed that we are still in a Quaternary interglacial stage between two glacial advances.

Secondary era — an obsolete name for the Mesozoic.

Silurian epoch — named after the Silures, an ancient Celtic tribe of the Welsh border-land, as proposed by Murchison, a British geologist, in 1835. The period extended from 435 to 395 million years ago, having a duration of 40 million years. It marks the final stage in the filling up of the Lower Paleozoic basis of deposition. At its top the first jawed fishes made their appearance. Trilobites (Arthropoda) were abundant and Brachiopoda were represented by all the major groups. Crinoids (Echinodermata) were present in sufficient numbers to form limestones, and cephalopods (Mollusca) were common. Lamellibranchs and gastropods (Mollusca) were locally abundant. The graptolites (Chordata) died out before the end of the period, and only in Central Europe did they continue into Lower Devonian times. The first land plants appeared in this epoch.

Tertiary system — the term used for the period of time which elapsed between the end of the Cretaceous epoch and the present time, lasting 65 million years — from 65 to 0 millions years ago — although the precise limits are defined variously by different authors. A division of the Tertiary may be made as follows: Holocene (youngest), Pleistocene, Pliocene, Miocene, Oligocene, Eocene, Paleocene (oldest). The term Tertiary (age of mammals) is traceable to Giovanni Arduina, who proposed the first geologic time-scale in 1760. As originally conceived, the name signified all relatively recent and more or less unconsolidated material that contained fossils resembling those still in existence. Geologists have thoroughly studied and subdivided the Tertiary because its rocks are young and almost everywhere are deposited over older rocks. Lyell's original classification (1833) included only the Eocene, Miocene, and Pliocene. The Oligocene was added in 1854 by E. Von Beyrech, and the Paleocene in 1874 by Carl Schimper. Lyell's idea of classifying on the basis of faunal comparison was excellent in theory, but later workers have had to rely on fossils other than those known to Lyell. During the Eocene there was a general increase in temperature over the Earth's surface, culminating in the upper Eocene Bartonian stage, which was characterized mainly by tropical and subtropical forms. The succeeding Oligocene showed a reversal of the conditions pertaining during the Eocene, with a general reduction in average temperature and regression of the seas. During Miocene and Pliocene times, the withdrawal of the seas continued. They are represented either by freshwater sediments in basins, or by the marine sediments deposited near existing coast lines. The gradual reduction in average temperature continued throughout this time.

Tillites = till — boulder clay.

Triassic epoch — a system named by Von Alberti, in 1934, after the threefold division of the Trias epoch in Germany. The period extends from 225 to 195 million years ago, a duration of 30 million years. It marks the beginning of the Mesozoic period and era. Widespread continental conditions which persist from the preceding Permian make the lower boundary difficult to interpret. The base of a 'zone' containing Lystrosaurus (a reptile) has been suggested but its general utility is doubtful. The upper limit of the Trias is marked by the sharp change in conditions at the start of the Jurassic. The only fossil remains of this time are those of reptiles (mainly as footprints). In Europe, where marine sediments occur, the fauna was largely one of lamellibranchs (Mollusca), crinoids (Echinodermata), and richly ornamented ammonoids (Mollusca). The earliest known dinosaur remains date from the Trias. Evaporites are found in the period and are of considerable economic importance.

BIBLIOGRAPHY

Investigational Papers on Health, Weather, and Climate Published by the Bioclimatology Unit of the Hadassah-Hebrew University Medical Center, Jerusalem.*

Joel, C. A., Meng, H., Parin, P., Selye, H., and Sulman, F. G., *Psyche & Hormon,* H. Huber, Bern, 1960.

Sulman, F. G. Bar-Joseph, N., and Hirschmann, N., Routine method for determination of urinary 17-hydroxycorticoids and its application in different diseases and in heat stress, *Isr. Med. J.,* 21, 220, 1962.

Joel, C. A., Meng, H., Parin, P., Selye, H., and Sulman, F. G., *Endocrinologia Psicosomatica,* Mongr., Ediciones Morata, Madrid, 1963.

Shanan, J., Brzezinski, A., Sharon, M., and Sulman, F. G., Active coping behaviour, anxiety and cortical steroid excretion in the prediction of transient amenorrhoea, *6th Eur. Conf. Psychosomat. Med.,* S. Karger, Basel, 1964.

Sulman, F. G., Hirschmann, N., and Pfeifer, Y., Effect of hot, dry desert winds (sirocco, sharav, hamsin) on the metabolism of hormones and minerals, *Proc. Lucknow Symp.on Arid Zones,* UNESCO, 1964, 89.

Shanan, J. Brzezinski,A., Sulman, F. G., and Sharon, M., Acting coping behavior,anxiety and cortical steroid excretion in the prediction of transient amenorrhea, *Behav. Science,* 10, 461, 1965.

Sulman, F. G., Wirkung des subtropischen Klimas auf die Arbeitsfaehigkeit, *Inform. Werksarzt Homburg,* 14, 154, 1967.

Koch, Y., Pfeifer, Y., and Sulman, F. G., Effect of climatic heat stress on the development of rats, *Int. J. Biometeorol.,*13, 93, 1969.

Danon, A., Weller, C. P., and Sulman, F.G., Mechanisms of reaction to heat stress, *Int. J. Biometeorol.,* 13, 95, 1969.

Weller, C. P., Dikstein, S., and Sulman, F. G., The effect of heat stress on body development in rats, Biometeorology (Suppl. to *Int. J. Biometeorol.,)*Vol.4 (Part II), 29, 1969.

Weller, C. P. and Sulman, F. G., Effect of climatic heat stress on catecholamine excretion, Biometeorology, (Suppl. to *Int. J. Biometeorol,)* Vol. 4 (Part II), 30, 1969.

Danon, A., Weller, C. P., and Sulman, F. G., Effect of dry hot desert winds on man, *Biometeorology,* (Suppl. to *Int. J. Biometeorol.,)* Vol. 4 (Part II), 71, 1969.

Danon, A. and Sulman, F. G., Ionizing effect of winds of ill repute on serotonin metabolism, Biometeorology, (Suppl.to *Int. J. Biometeorol.,)*Vol. 4 (Part II),135, 1969.

Sulman, F. G., Effect of heat stress on release of Catecholamines, serotonin and other hormones, Abstr. IV, *Int. Congr. on Pharmacol.,* Schwabe, Basel, 1969, 395.

Sulman, F. G., Danon, A., Pfeifer, Y., Tal, E., and Weller, C.P., Urinalysis of patients suffering from climatic heat stress (sharav), *Int. J. Biometeorol.,* 14, 45, 1970.

Dikstein, S., Kaplanski, Y., Koch, Y., and Sulman, F. G., The effect of heat stress on body development of rats, *Life Sci.,* 9, 1191, 1970.

Sulman, F. G., Meteorologische Frontverschiebung und Wetterfuehligkeit — Foehn, Chamssin, Scharaw, *Aerztl. Praxis,* 23, 998, 1971.

Sulman, F. G., Serotonin-migraine in climatic heat stress, its prophylaxis and treatment, *Proc. Int. Headache Symp.,* "Headache" Suppl., Elsinore, Denmark, 1971, 205.

Tal, E., Superstine, E., and Sulman, F. G., Effect of heat on serum thyroxin and thyrotropin and its modification by dihydrotachysterol *Experientia,* 27, 1299, 1971.

Dikstein, S. and Sulman, F. G., Prevention of cold stress by anabolic agents, *Isr. J. Med. Sci.,* 8, 572, 1972.

Tal, E. and Sulman, F. G., Effect of dihydrotachysterol on TSH secretion in rats, *Neuroendocrinology,* 9, 142, 1972.

Tal, E. and Sulman, F. G., Urinary thyroxine test, *Lancet,* 1, 1291, 1972.

Sulman, F. G. and Superstine, E., Ageing and adrenal-medulla exhaustion due to lack of monoamines and raised monoamine-oxidase levels, *Lancet,* 2, 663, 1972.

Sulman, F. G., Urinalysis and treatment of patients suffering from climatic heat stress (sharav), *Pediatric Work Physiol., Proc. 4th Int. Symp., Wingate Inst. Israel,* 4, 335, 1972.

Tal, E. and Sulman, F. G., Rat thyrotrophin levels during heat stress and stimulation by thyrotrophin releasing factor, *J. Endocrinol.,* 57, 181, 1973.

Tal, E. and Sulman, F. G., Dehydroepiandrosterone-induced thyrotrophin release in immature rats, *J. Endocrinol.,* 57, 183, 1973.

Sulman, F. G., Pfeifer, Y., and Superstine, E., Adrenal medullary exhaustion from tropical winds and its management, *Isr. J. Med. Sci.,* 8, 1022, 1973.

* In chronological order.

Sulman, F. G., Management of intractable migraine by a combination of pizotifen (Sandomigran) with methylergotamine (My 25-Sandoz), *10th Int. Congr. Neurology,* Excerpta Medica, Barcelona, 1973, No. 754, 237.

Pfeifer, Y. and Sulman, F. G., Ionometry of hot dry winds (khamsin, sharav) and its application to ionizing treatment of weather-sensitive patients, *Isr. J. Med. Sci.,* 9, 686, 1973.

Sulman, F. G. and Tal, E., Treatment of functonal hypothyroidism by oral TRF monitored by daily urinary thyroxine and histamine assay, *Horm. Metab. Res.,* 6, 92, 1974.

Sulman, F. G., Dikstein, S., Hirschmann, N., Kaplanski, Y., Koch, Y., Nir, I., Pfeifer, Y., Superstine, E., Tal, E., and Weller, C. P., Effect of ambient heat stress on body development of rats and survival of fatal heat stress by drug administration, *Proc. 2nd Symp. Temperature Regulation and Drug Action,* S. Karger, Paris, 1974, 339.

Sulman, F. G., Sensibilita dell' uomo alle variazioni dei fronti meteorologici, *Gazzetta San.,* 45, 10, 1974.

Sulman, F. G., Pfeifer, Y., Shalita, B., and Tal, E., Air sterilisation: influence of negative ionisation on bacterial counts on agar plates exposed to air, *Int. Res. Comm. Syst.,* 2, 1452, 1974.

Sulman, F. G., Assael, M., Alpern, S., and Pfeifer, Y., Influence of artificial ionization of air of the electroencephalogram, *Isr. J. Med. Sci.,* 10, 568, 1974.

Sulman, F. G., Bioklimatologie trocken-heisser Winde, *Promet-Meteorologische Fortbildung,* (Ed. Deutscher Wetterdienst, Frankfurt) 4, 17, 1974.

Sulman, F. G., Urinalysis and treatment of patients suffering from climatic heat stress (sharav), *Medic (Monthly Drug Compilation, Israel),* 3, 26, 1974.

Sulman, F. G., Climatic factors in the incidence of attacks of migraine, *Hemicrania,* 6, 2, 1974.

Sulman, F. G., Meteorological front movements and human weather sensitivity, *Karger Gazette,* 30, 1, 1974.

Sulman, F. G., Foehnleiden, ihre Ursachen und Behandlung, *Phys. Med. Rehab.,* 15, 256, 1974.

Assael, M., Pfeifer, Y., Sulman, F.G., Alpern, S., and Shalita, B., Influence of artificial air ionisation on the human electroencephalogram, *Int. J. Biometeorol.,* 18, 306, 1974.

Sulman, F. G., Levy, D., Lewy, A., Pfeifer, Y., Superstine, E., and Tal, E., Air ionometry of hot, dry desert winds (sharav) and treatment with air ions of weather-sensitive subjects, *Int. J. Biometeorol.* 18, 383, 1974.

Sulman, F. G., Pfeifer, Y., and Superstine, E., Disturbances of homeostasis by heat stress or aging and its treatment with minidoses of MAO blockers, *5th Int. Congr. Hormones, Homeostasis and Brain,* Elsevier, Amsterdam, 1974, 37.

Sulman, F. G., Tal, E., Pfeifer, Y., and Superstine, E., Intermittent hyperthyreosis — a heat stress syndrome, *Horm. Metabol. Res.,* 7, 424, 1975.

Sulman, F. G., Levy, D., Pfeifer, Y., Superstine, E., and Tal, E., Effect of natural and artificial air ionisation on urinary neurohormone excretion in man, *Int. J. Biometeorology,* 19, 202, 1976.

Tal, E. and Sulman, F. G., Dehydroepiandrosterone-induced thyrotrophin release during heat stress in rats, *J. Endocrinol.,* 67, 99, 1975.

Tal, E., Pfeifer, Y., and Sulman, F. G., Effect of ionization on blood serotonin in vitro, *Experientia,* 32, 326, 1976.

Sulman, F. G., *Health, Weather and Climate,* Monogr., S. Karger, Basel, 1976.

Koren, E., Wapnik, S., Solowiejczyk, M., Pfeifer, Y., and Sulman, F. G., Serotonin in the portal vein after acidification, *Int. Surg.,* 61, 370, 1976.

Sulman, F. G., Doping and sports, in *Physical Education and Medicine,* Jerusalem Academy of Med. Publ., Jerusalem, 1976, 27.

Sulman, F. G., Effect of ambient heat stress on body development of rats and effect of drugs on heat stress survival, Congr. Rep., *Isr. J. Med. Sci.,* 12, 1134, 1976.

Tal, E., Chayoth, R., Zor, U., Goldhaber, G., and Zerachie, A., Pituitary-thyroid interrelationship in rats exposed to different environmental temperatures, *Acta Endocrin.,* 83, 99, 1976.

Sulman, F. G., Pfeifer, Y., and Tal, E., Migraene-Behandlung durch Enzyminduktion mit Proxibarbal, *Ther. G.,* 115, 2088, 1976.

Sulman, F. G., Levy, D., and Lunkan, L., Wetterfuehligkeit und ihre Beziehung zu Sferics, Ionen und Electrofeldern, *Z. Phys. Med.,* 5, 229, 1976.

Sulman, F. G., Pfeifer, Y., and Tal, E., Effect of enzyme induction by barbiturates on neurohormone excretion in man, *Isr. J. Med. Sci.,* 12, 1521, 1976.

Sulman, F. G., Tal, E., Pfeifer, Y., and Superstine, E., Intermittent hyperthyreosis — a clinical heat stress syndrome, in *Drugs, Biogenic Amines and Body Temperature,* Cooper, K. E., Lomax, P.,and Schoenbaum, E., Eds., S. Karger, Basel, 1977, 181.

Sulman, F. G., Levy, D., Lunkan, L., Pfeifer, Y., and Tal, E., Behandlung der Wetterfiihligkeit, *Fortsch. Med.,* 95, 746, 1977.

Sulman, F.G. Prevention of allergy by enzyme induction, *Lancet,* 1, 1206, 1977.

Sulman, F. G., Levy, D., Lunkan, L., Pfeifer, Y., and Tal, E., Human reactions to climatic fluctuations, *Int. Conf. Meteorol. of Semi-Arid Zones,* Government Meteorological Service, Tel Aviv, 1977, 54.

Sulman, F. G., Pfeifer, Y., Levy, D., Lunkan, L., and Superstine, E., Human weather sensitivity and atmospheric electricity, *Israel Meteorological Research Papers,* Steinitz Memorial Vol. 1, 1977, 42.

Sulman, F. G., Levy, D., Pfeifer, Y., Superstine, E., and Tal, E., Sensitivity to hot-dry sirocco winds and its treatment by air ionisation, *La Clinica Termale,* 29, 1, 1976.

Sulman, F. G., Wetterfuehligkeit und Wetterempfindlichkeit, *Hexagon Rubrik (Basel),* 5, I, 1977.

Sulman, F. G., Meteorosensibilitetet meteoropathie, *Hexagone Suppl. (Basel),* 5, I, 1977.

Sulman, F. G., Weather sensitivity, its diagnosis and treatment, *Hexagon Suppl. (Basel),* 5, I, 1977.

Sulman, F. G., Pfeifer, Y., and Superstine, E., The adrenal exhaustion syndrome: an adrenal deficiency in long-distance runners, *Ann. N.Y. Acad. Sciences,* 301, 918, 1977.

Sulman, F. G., Pfeifer, Y., and Superstine, E., Preventive treatment of migraine with mini-doses of dani-tracene, *Headache,* 17, 203, 1977.

Sulman, F. G., Levy, D., Lunkan, L., Pfeifer, Y., and Tal, E., Absence of harmful effects of protracted negative air ionisation, *Int. J. Biometeorol.,* 22, 53, 1978.

Sulman, F. G., Vaerfolsomhet og vaeroverfolsomhet diagnosa og behandling, (Swedish), *Hexagon Rubrik (Basel),* 5, I, 1978.

Sulman, F.G., Meteorosensibilidad & meteoropatia, Diagnostico y tratamiento, (Spanish), *Hexagono Suplemento,* 5, I, 1978.

Pfeifer, Y., Tal, E., and Sulman, F. G., Sharav-induced migraine and its treatment, *Harefuah (J. Isr. Med. Assoc.),* 95, 158, 1978.

Sulman, F. G. and Superstine, E., The syndrome of adrenal exhaustion in geriatrics and its management (Hebrew), *Georontologia,* 12, 33, 1978.

Sulman, F. G., Weergevoeligheid — diagnose en therapie, (Swedish), *Hexagon Rubriek,* 1, I, 1978.

Sulman, F. G., Aerionotherapy, in *Wholistic Dimensions in Healing: A Resource Guide,* Kaslof, L. J., Ed., Doubleday, Garden City, N.Y., 1978, 139.

Sulman, F. G., Bioclimatology and Ionotherapy, in *Electro-Vibratory Body,* Beasley, V. R., Ed., University of Trees Press, Boulder Creek, Calif., 1978, 117.

Behar, A. J., Deutsch, E., Pomerantz, E., Pfeifer, Y., and Sulman, F. G., Migraine, serotonin and the carotid body, *Lancet,* 1, 550, 1979.

Levy, D., Lunkan, L., and Sulman, F. G., Weather sensitivity and atmospheric electricity, *Harafuah (J. Isr. Med.Assoc.),* 95, 414, 1978/79.

Tannenbaum, J., Pfeifer,Y., Tal, E., and Sulman, F. G., Effect of cerebral electro-therapy on neurohormone excretion in weather-sensitive patients, *Int. J. Biometeorol.,* 23, 131, 1979.

Sulman, F. G., Pfeifer, Y., and Goldgraber, M. B., Treatment of multiple allergyby enzyme induction with proxibarbal, X. *Int. Congr. Allergology,* Jerusalem, 1979, 316.

Sulman, F. G., Meteorosensibility and meteoropathy, (Greek translation), *Hexagon Suppl.,*3(2), 1, 1979.

Sulman, F. G., Reactions to hot dry desert winds and their treatment, *Harefuah (J. Isr. Med. Assoc.),*96, 185, 1979.

Sulman, F. G., Neue Methoden zur Behandlung der Migraene, *Cytobiol. Rev.,* 3, 108, 1979.

Sulman, F. G., Negative ion therapy, in *Alternative Medicine,* Stanway, A., Ed., McDonald and Jane's Publ., London, 1979, 108.

Sulman, F. G., Gonadotropi ormoni, *Enciclopedia Medica Italiana,* 7, 559, 1979.

Sulman, F. G., *The Effect of Air Ionization, Electric Fields, Atmospherics and other Electric Phenomena on Man and Animal,* Monogr., Charles C Thomas, Springfield, Ill., 1980.

Sulman, F. G., Wetterbeschwerden, Klimabedingte Leiden und Erkrankungen, ihre Ursachen und ihre Behandlung, *Medizinische Ratgeber,* Monogr., Koehnlechner, M. Ed., Heyne Publ., Munich, 1980.

Sulman, F. G., Pfeifer, Y., and Superstine, E., Preventive treatment of migraine with enzyme induction by proxibarbal in a "double-blind" trial, *Headache,* 20, 269, 1980.

Sulman, F. G., Pfeifer, Y., and Superstine, E., Preventive treatment of serotonin-migraine with Org. CC 94. A double-blind study, *Arzneimittel-Forschung (Drug Research),* 31-I, 109, 1981.

Sulman, F. G., Pfeifer, Y., and Tal, E., Effects of hyperthyroidism on urinary histamine and creatinine, *Isr. J. Med. Sci.,* 16, 621, 1980.

Sulman, F. G., Die Wetteremofindlichkeit und die neuen medizinischen Methoden ihrer Behandlung, *Universitas,* 36, 43, 1981.

Sulman, F. G., The impact of atmospheric electricity of building designs, *9th Int. Congr. Biometeorology,* Osnabrueck CHP-17, Abstr. 1981, 81.

Merimsky, E., Litmanovitch, Y. I., and Sulman, F. G., Prevention of post-operative thromboembolism by negative air ionization in a double-blind study, *9th Int. Congr. Biometeorology,* Osnabrueck CHP-18, Abstr. 1981, 82.

Sulman, F. G., Twelve sources of air electricity and their impact on man, *9th Int. Congr. Biometeorology,* Osnabrueck, CTS-04-PO, Abstr. 1981, 278.

INDEX

Papagayo, 25
Parathyroid glands, 152
Particulate matter in atmosphere, 140—143
Permian epoch, 156
Personality changes, 1
Physiological effects of negative ions, 13
Pineal gland, 152
Pines (*Pinus aristata*), 85
Pituitary, 152
Pizotifen, 42
Planning of towns, 57
Pleistocene period, 106
  defined, 156
Pliocene epoch, 77, 106
  defined, 156
Polar ice masses, 122—123
Pole wandering, 99, 102—103
  defined, 156
Pollen rain, 89
Pollen zone dating, 87—91
Pollution
  of stratosphere, 143
  sulfur dioxide, 141
Postglacial climatic optimum, 115
Postglacial times, 107—109, 115—116
Postthermal syndrome, 8
Potassium, 39
  excess of, 36—37
Potassium-argon dating, 83—84
Power generation, 63—64
Precambrian epoch, 156
Precession, 125
Prehistoric climate, 112—113
Primary era, 157
Prolactin, 5
Protactinium-ionium dating, 83
Proterozoic period, 73
  defined, 157
Proton effect, 131—132
Psychosomatic disorders, 10—13
Psychotechnic disorders, 10—13
Ptolemy, 79
Pulsars, 126—127
Pulsating stars, 126—127
Pyrnwind, 25

Q

Quaternary era, 77—78
  climatic changes during, 103—104
  defined, 157

R

Radiation, 127—133
Radiocarbon dating, see Carbon 14 dating
Radiometric dating, 80—84
Rail transport, 68—69
Rain, 89
Recreation, 18

Redwoods (*Sequoia*), 54—55, 85
Relative Humidity (RH), 150
Reshabar, 25, 34
Reticular formation, 152
Reversals in geomagnetic field, 94
RH, see Relative humidity
Rhizocarpon geographicum, 86—87
River diversion, 144
Road transport, 69
Roman calendar, 79
Roteturmwind, 25
Rotor, 150
Routing of weather, 67
Rubidium-strontium dating, 84

S

Sansar, 25
Santa Ana, 21—24, 34
  criminality and, 23—24
Savanna, 150
Sea level changes, 119—120
Secondary climatic optimum, 108—109
  of Middle Ages, 115—117
Secondary era, 157
Sedimentation dating, 93—94
Seistan, 34
Sensitivity to weather, 38—50
Sequoia (giant redwoods), 54—55, 85
Serotonin, 4, 15, 39
  defined, 152
  hyperproduction of, 8
  ionization and, 38
  release of, 38, 40—42
Serotonin stress reaction, 3—4
Sex hormones, 5
Sferics, 38
Shamal, 34
Sharav, 32—38
Sharkiye, 32, 34
Sharqi, 34
Sheep, 56
Shopping centers, 71
Silurian epoch, 73
  defined, 157
Simoon, 34
Simum, 34
Sirocco, 25, 32, 34
Smog, 150
Sodium, 39
  loss of, 36—37
Solar houses, 65
Solar ionization, 132
Solar magnetism, 130—131
Solar radiation, 127—133
Solar spots, see Sunspots
Southerly Burster, 24
Southerly Buster, 24
Southern Foehn, 25—28
Sphagnum, 91
Spoerer Minimum, 130